JOURNEYMAN WIREMAN

By Tom Henry

Illustrations by Mary Elizabeth

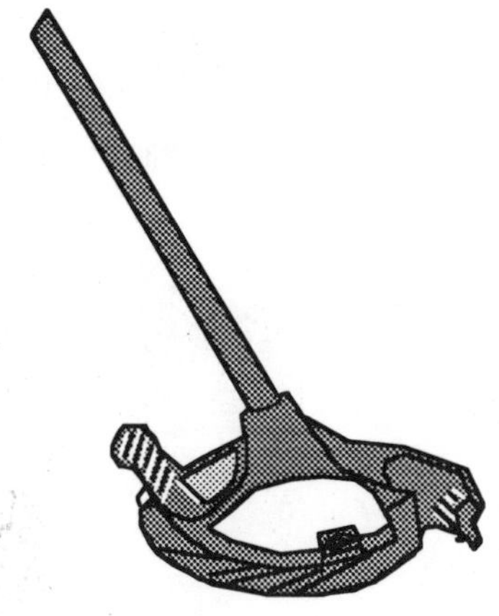

ISBN 0 - 945495 - 19 - 6

Preface

This book was written to help prepare an electrician for the mechanical aptitude test which many areas include in the electrical license examination.

Every electrician should be familiar with the tools of his profession. Tool recognition quizzes have become important in mechanical ability, comprehension and aptitude testing.

Most electricians learn the electrical profession through an apprenticeship program. I was fortunate to have served a four year IBEW apprenticeship starting in 1956.

However, some electricians have learned the trade by working as a helper for many years, or through a vocational trade school, or by a correspondence course. Some lack the mathematics, blue print reading, formal code classes, etc.

Proper wiring methods, installation, tool identification, reading blueprints, switch connections, motor control, circuit testing, meter reading, etc. are some of the areas covered in this preparation book.

This book contains 62 Journeyman quizzes plus two 3-hour Journeyman exams. Most of the quizzes are from pictures. This is the part I really like. Many books contain all text which becomes boring to the working electrician.

Even the answer section is done in the same pictures to help the electrician in preparing for the identification part of the exam.

The book starts off with quizzes of the type of "fill in the blank", which are the toughest. I have also included "true or false" or "multiple choice" type.

The 3-hour exam includes the parts board identification which is a required test in some areas of the USA. I'm in favor of this type testing, as the Journeyman should be able to identify the electrical equipment that is being installed or maintained.

If you have trouble with the Ohms law - theory or the Code portion of the exam it is highly recommended that you review other books that I have written for the Journeyman exam.

Be properly prepared before taking any exam and you'll only have to take it once!

Tom Henry

JOURNEYMAN WIREMAN QUIZ #1
BLUEPRINT SYMBOLS

•Fill in the blank with the correct name for the symbol

1. WH ____________________

2. ____________________

3. ____________________

4. ____________________

5. ____________________

6. ____________________

7. S_F ____________________

8. ____________________

9. M ____________________

10. ____________________

JOURNEYMAN WIREMAN QUIZ #2
SWITCH QUESTIONS

• *Circle the correct answer.*

1. Which of the following is properly connected?

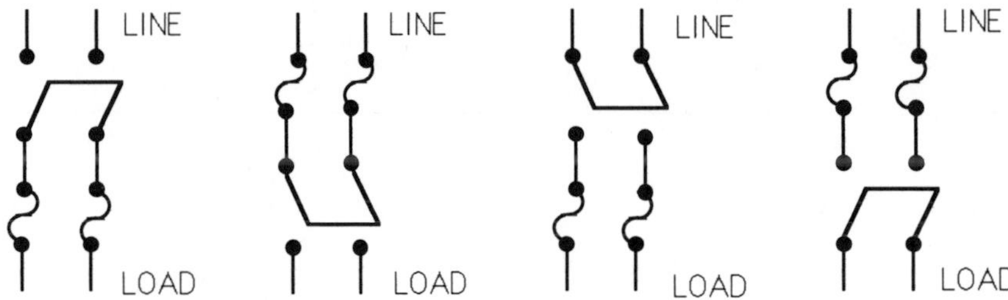

2. Which of the following double-pole double-throw switches is properly connected as a reversing switch?

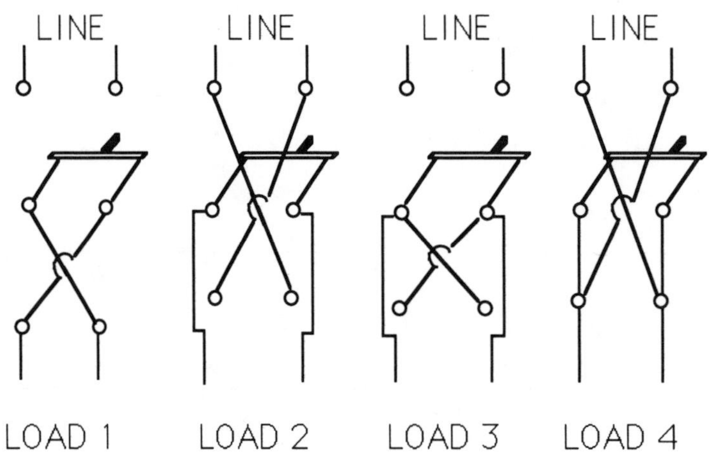

JOURNEYMAN WIREMAN QUIZ #3
TOOL IDENTIFICATION QUESTIONS

• *Fill in the blank with the correct name for the tool.*

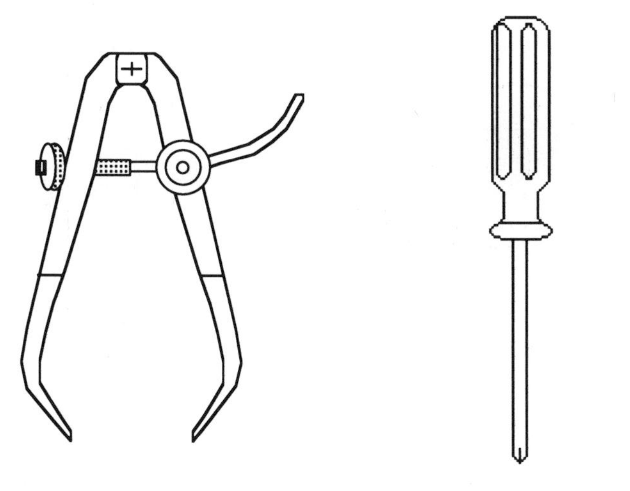

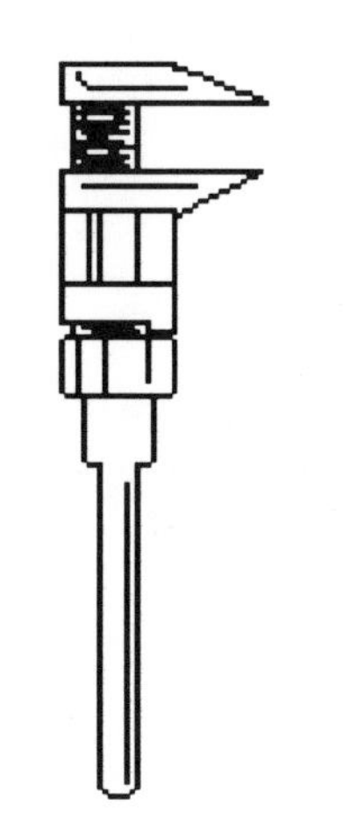

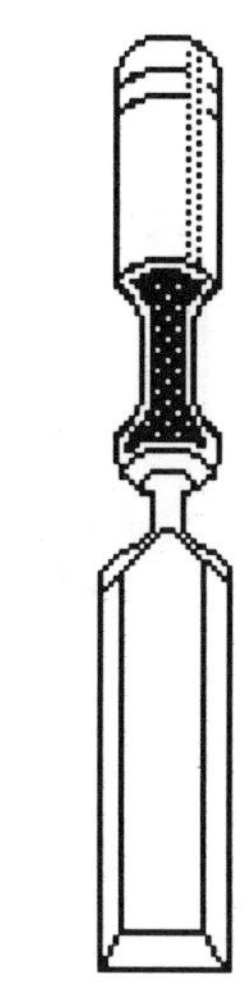

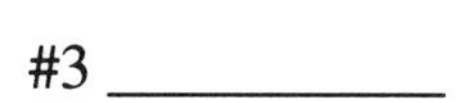

#1 ____________ #2 ____________ #3 ____________ #4 ____________

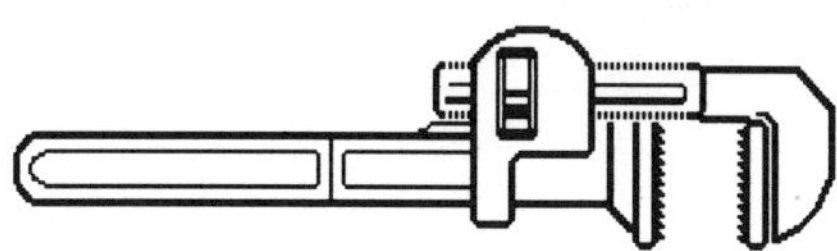

#5 ____________ #6 ____________

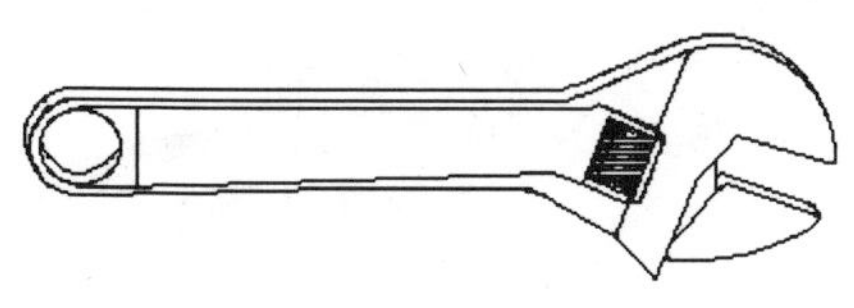

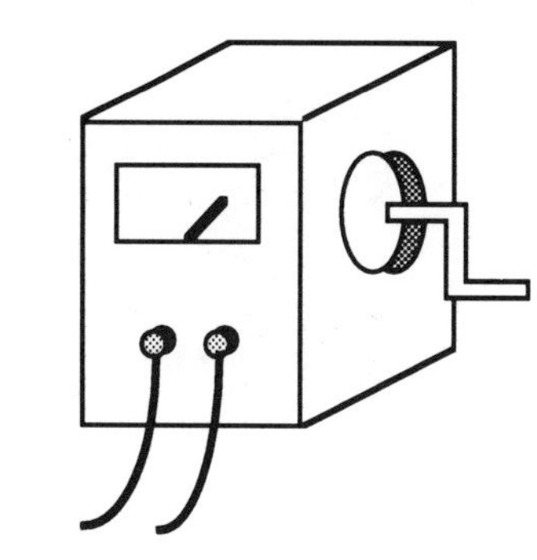

#7 ____________ #8 ____________

JOURNEYMAN WIREMAN QUIZ #4
WIRING METHODS

• *Fill in the blank with the correct name for the wiring method or equipment.*

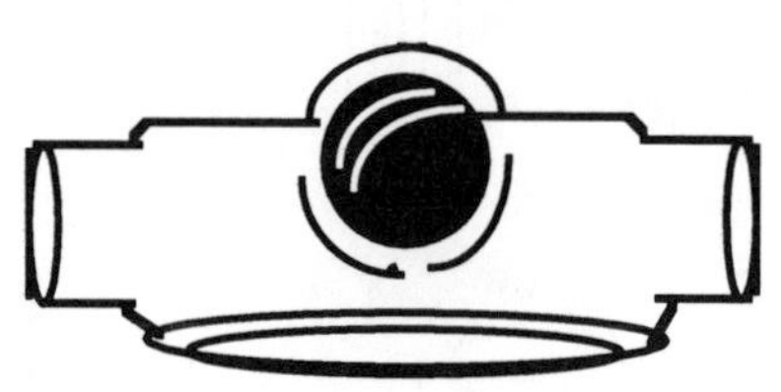

1. ____________________
2. ____________________
3. ____________________

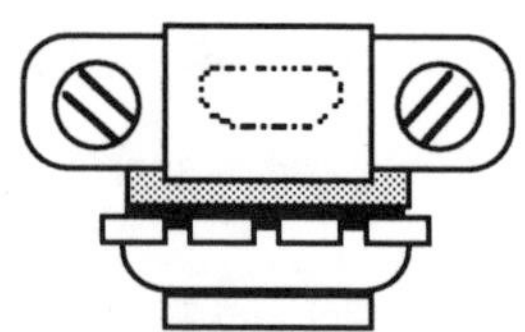

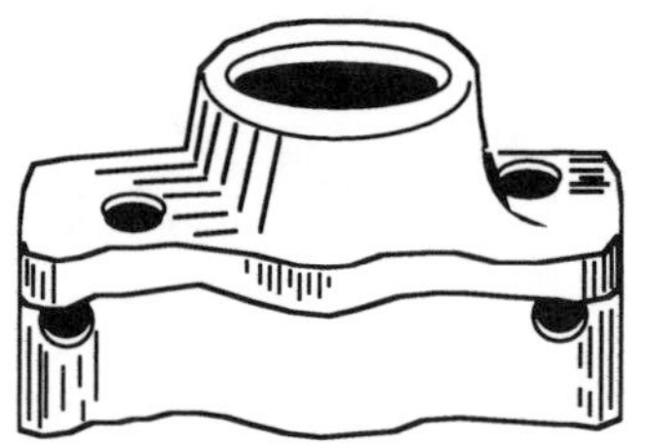

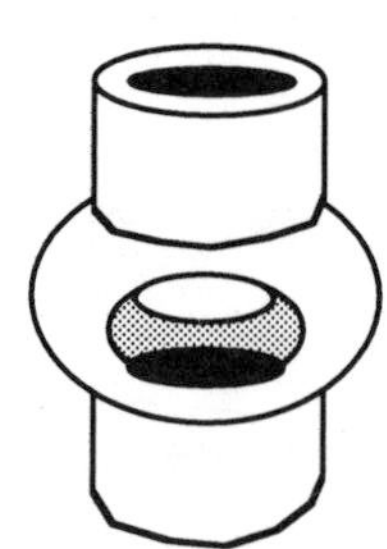

4. ____________________
5. ____________________
6. ____________________

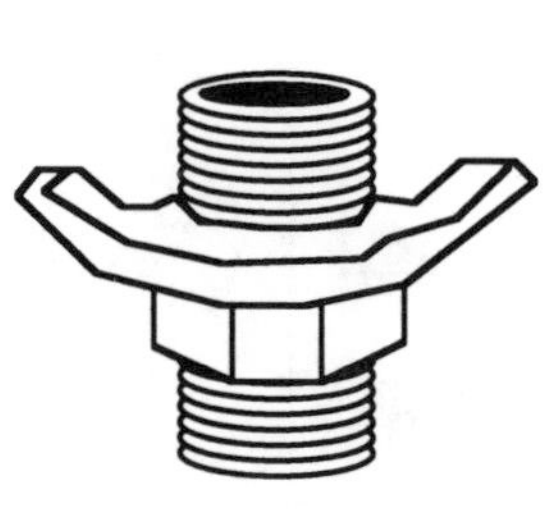

7. ____________________
8. ____________________

JOURNEYMAN WIREMAN QUIZ #5
PRACTICAL QUESTIONS

•*Circle the correct answer:*

1. A switch is a device for _____.

I. making or breaking connections
II. changing connections
III. interruption of circuit under short-circuit conditions

(a) I only (b) I and II only (c) II and III only (d) I, II and III

2. One of the essential functions of any switch is to maintain a _____.

(a) good high-resistance contact in the closed position
(b) good low-resistance contact in the closed position
(c) good low-resistance contact in the open position
(d) none of these

3. For a given line voltage, four heater coils will consume the most power when connected _____.

(a) all in series
(b) all in parallel
(c) with 2 parallel pairs in series
(d) 1 pair in parallel with the other two in series

4. All edges that are invisible should be represented in a drawing by lines that are _____.

(a) dotted
(b) broken
(c) curved
(d) solid

5. Two switches in one box under one face-plate is called a _____.

(a) double-pole switch
(b) two-gang switch
(c) 2-way switch
(d) 4-way switch

JOURNEYMAN WIREMAN QUIZ #6
BLUEPRINT SYMBOLS

•Fill in the blank with the correct name for the symbol

1. 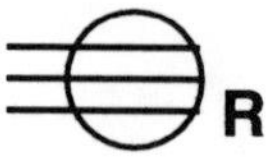 R ______________________________

2. ______________________________

3. ______________________________

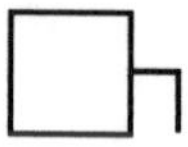

4. ______________________________

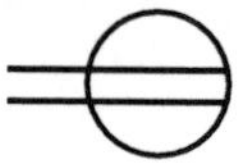

5.  D ______________________________

6. ______________________________

7.  H ______________________________

8. S_3 ______________________________

9. ______________________________

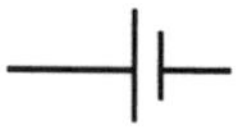

10.  J ______________________________

JOURNEYMAN WIREMAN QUIZ #7
TOOL IDENTIFICATION QUESTIONS

• *Fill in the blank with the correct name for the tool.*

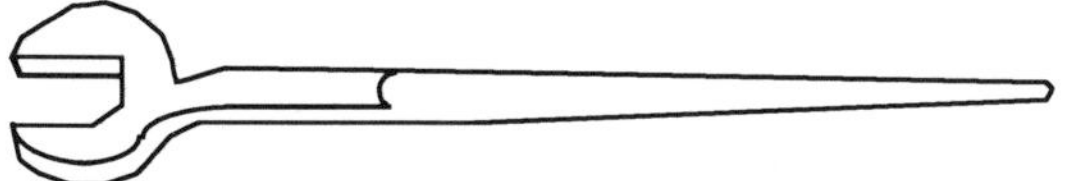

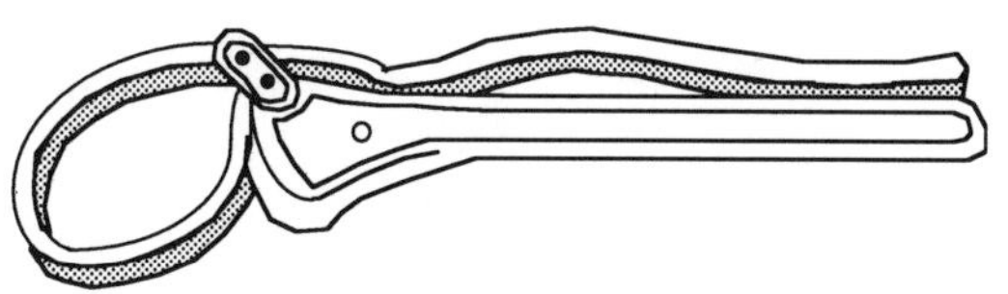

1. ______________________________

2. 

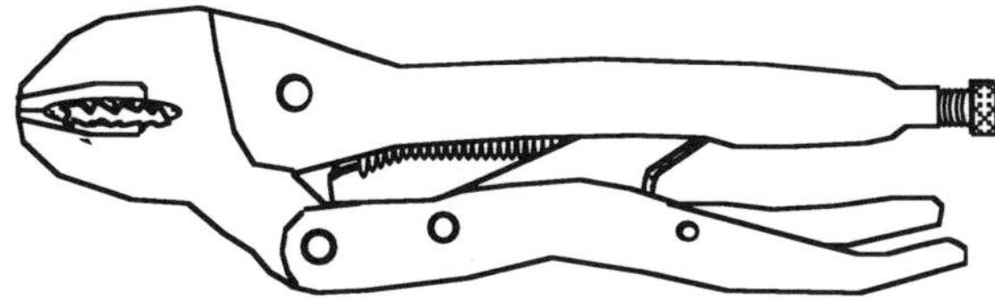

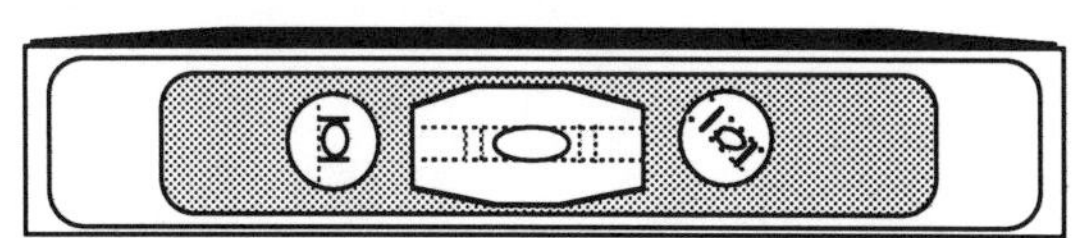

3. ______________________________

4. 

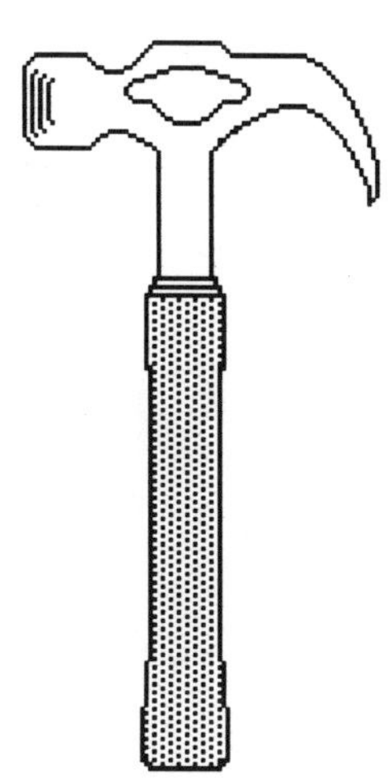

#5 ____________ #6 ____________ #7 ____________

JOURNEYMAN WIREMAN QUIZ #8
CONNECTIONS

1. Connect the following three single-phase transformers delta-wye three-phase.

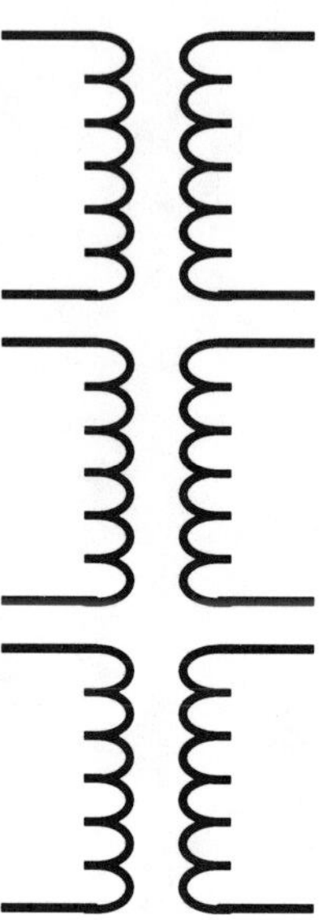

2. Which of the following is the correct wiring to a light controlled by two 3-way switches?

- *Circle the correct diagram.*

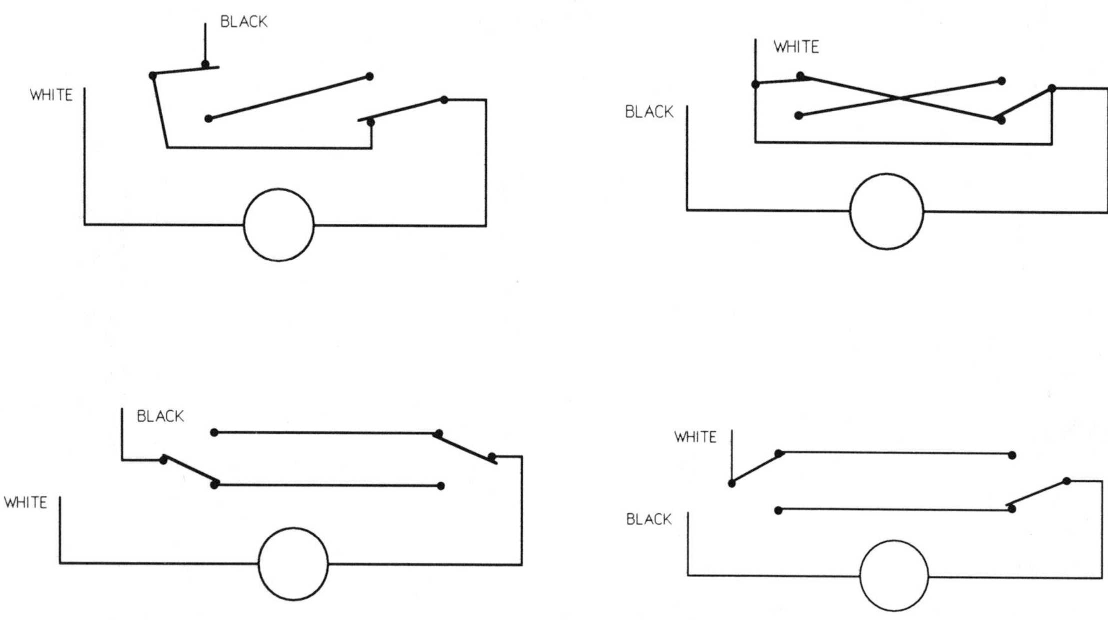

JOURNEYMAN WIREMAN QUIZ #9
PRACTICAL QUESTIONS

•*Circle the correct answer:*

1. When a gauge number such as "No.4" is used in connection with a twist drill, it refers to the ____.

(a) length
(b) hardness
(c) number of cutting edges
(d) diameter

2. Galvanized conduit has a finish exterior and interior of ____.

(a) lead
(b) copper
(c) zinc
(d) nickel

3. When stripping insulation from an aluminum conductor ____.

I. remove insulation as you would sharpen a pencil
II. ring the conductor and slip the insulation off the conductor
III. peel the insulation back and then cut outwards

(a) I, II and III (b) I and II only (c) I and III only (d) II and III only

4. A ____ is used to test the electrolyte of a battery.

(a) growler
(b) hydrometer
(c) manometer
(d) voltmeter

5. A drawing showing the floor arrangement of a building is referred to as a(an) ____.

(a) perspective
(b) isometric
(c) surface G.B.
(d) plan

JOURNEYMAN WIREMAN QUIZ #10
CONTROLS

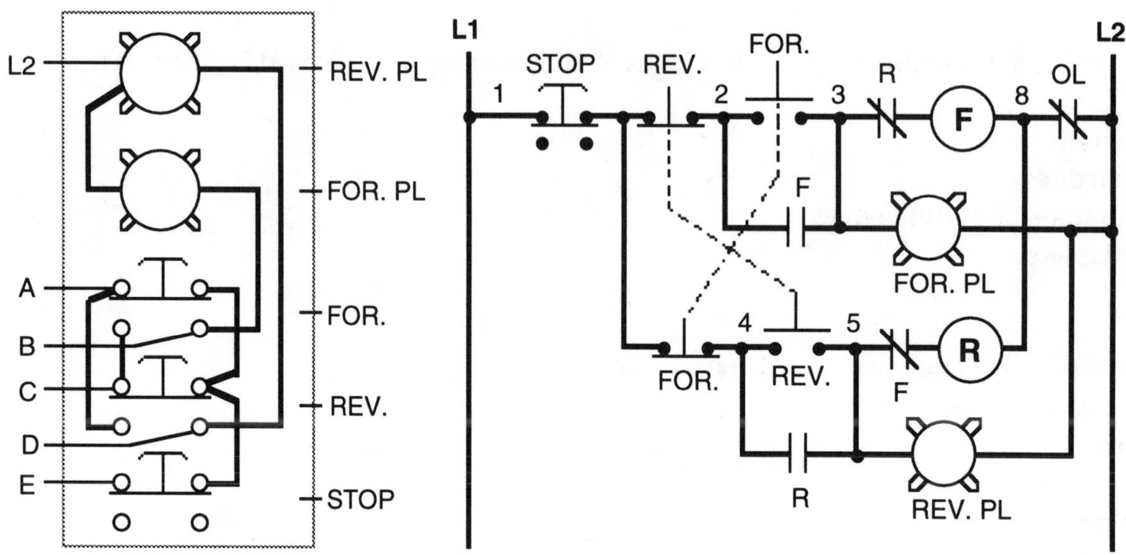

1. Conductor "A" on the push button station is conductor number ____ shown in the control circuit schematic.

(a) 1 (b) 2 (c) 3 (d) 4 (e) 5

2. Conductor "B" on the push button station is conductor number ____ shown in the control circuit schematic.

(a) 1 (b) 2 (c) 3 (d) 4 (e) 5

3. Conductor "C" on the push button station is conductor number ____ shown in the control circuit schematic.

(a) 1 (b) 2 (c) 3 (d) 4 (e) 5

4. Conductor "D" on the push button station is conductor number ____ shown in the control circuit schematic.

(a) 1 (b) 2 (c) 3 (d) 4 (e) 5

5. Conductor "E" on the push button station is conductor number ____ shown in the control circuit schematic.

(a) 1 (b) 2 (c) 3 (d) 4 (e) 5

JOURNEYMAN WIREMAN QUIZ #11
TOOL IDENTIFICATION QUESTIONS

• *Fill in the blank with the correct name for the tool.*

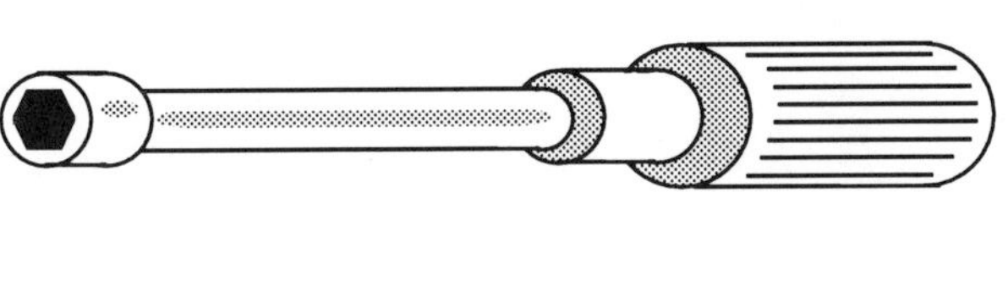

1. ______________________________

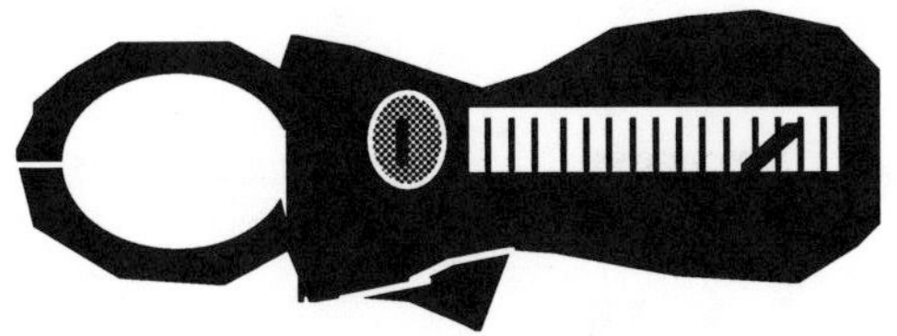

2. ______________________________

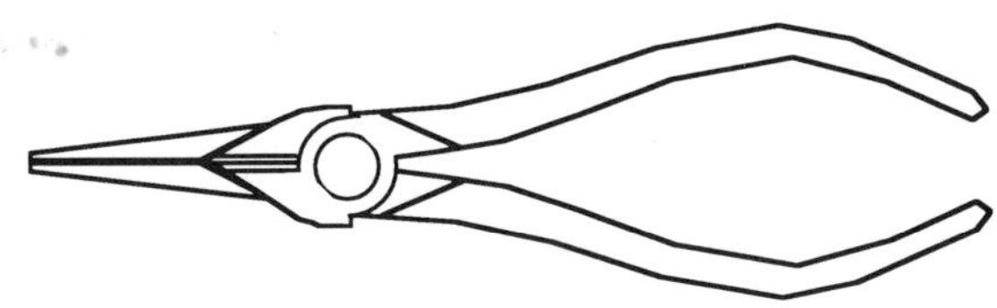

3. ______________________________

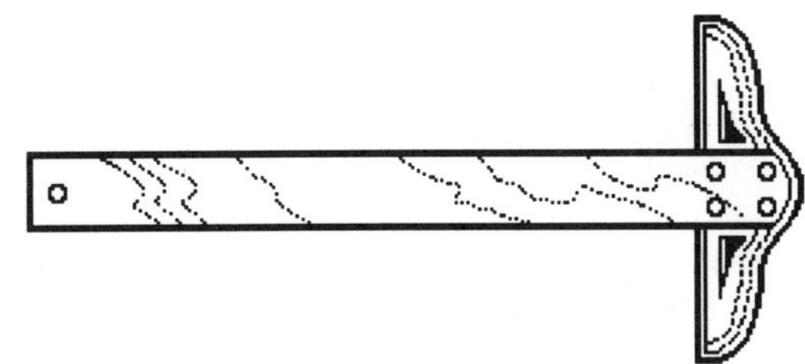

4. ______________________________

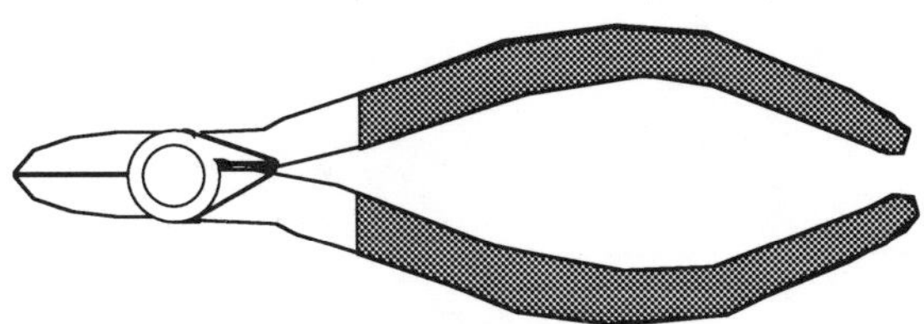

5. ______________________________

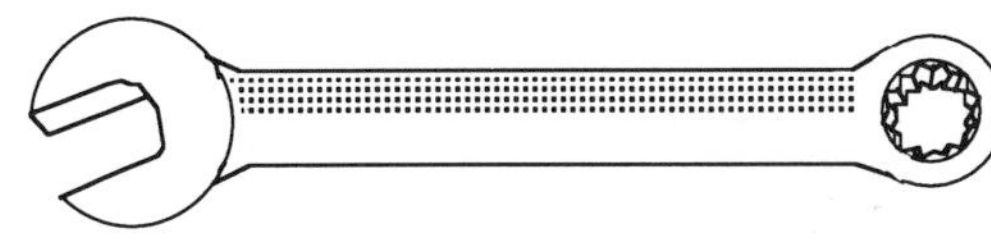

6. ______________________________

JOURNEYMAN WIREMAN QUIZ #12
BLUEPRINT SYMBOLS

•Fill in the blank with the correct name for the symbol

1. S_K ____________________

2. ____________________

3. F ____________________

4. TP ____________________

5. R ____________________

6. ____________________

7. T ____________________

8. ____________________

9. WP ____________________

10. J ____________________

JOURNEYMAN WIREMAN QUIZ #13
PRACTICAL QUESTIONS

•*Circle the correct answer:*

1. The purpose of a clip clamp is to ____.

I. ensure good contact between the fuse terminals of cartridge fuses and the fuse clips
II. make it possible to use cartridge fuses of a smaller size than that for which the fuse clips are intended
III. prevent the accidental removal of the fuse due to vibration

(a) I, II and III (b) I only (c) II only (d) I and II only

2. An electric bell outfit would be used to check for ____.

(a) voltage
(b) ampacity
(c) continuity
(d) current

3. A hickey is ____.

(a) a tool used to bend small sizes of rigid conduit
(b) a part of a conduit
(c) not used in the electrical trade
(d) used only by a plumber

4. When using a #12-2 with ground cable, the ground ____ carry current under normal operation.

(a) will
(b) will not
(c) will sometimes
(d) none of these

5. The load side is usually wired to the blades of a knife switch to ____.

(a) prevent blowing the fuse when opening the switch
(b) make the blades dead when the switch is opened
(c) prevent arcing when the switch is opened
(d) allow changing of fuses without opening the switch

JOURNEYMAN WIREMAN QUIZ #14
WIRING METHODS

• *Fill in the blank with the correct name for the wiring method or equipment.*

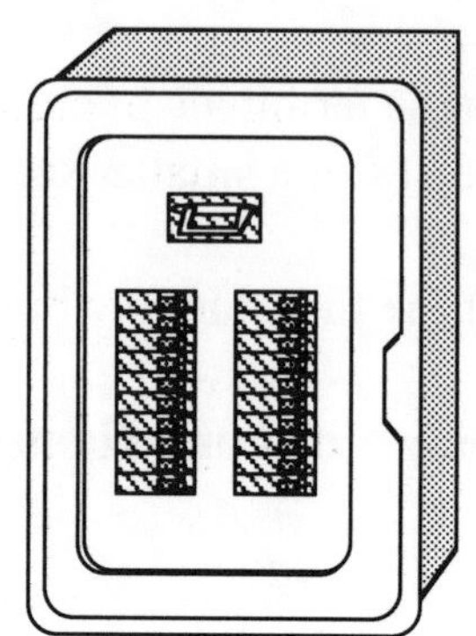

1. ____________________

2. ____________________

3. ____________________

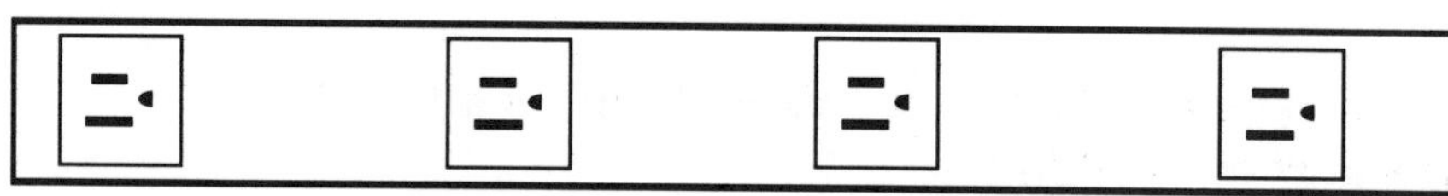

4. ____________________

5. ____________________

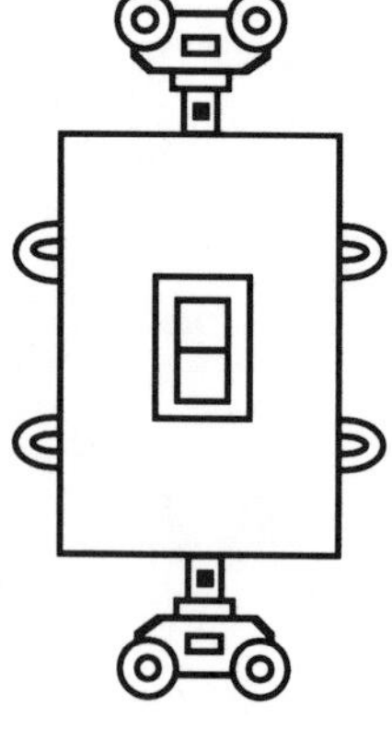

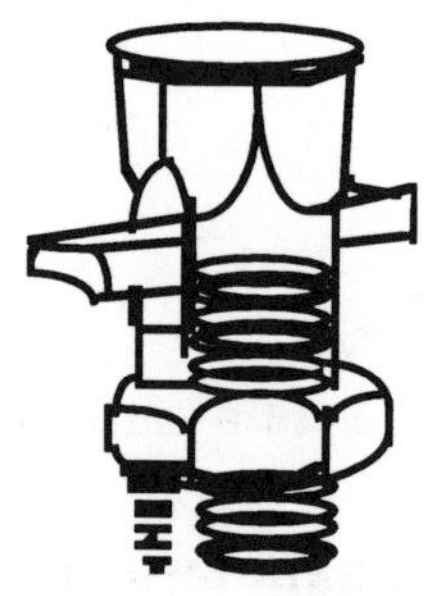

6. ____________________

7. ____________________

8. ____________________

JOURNEYMAN WIREMAN QUIZ #15
TOOL IDENTIFICATION QUESTIONS

• *Fill in the blank with the correct name for the tool.*

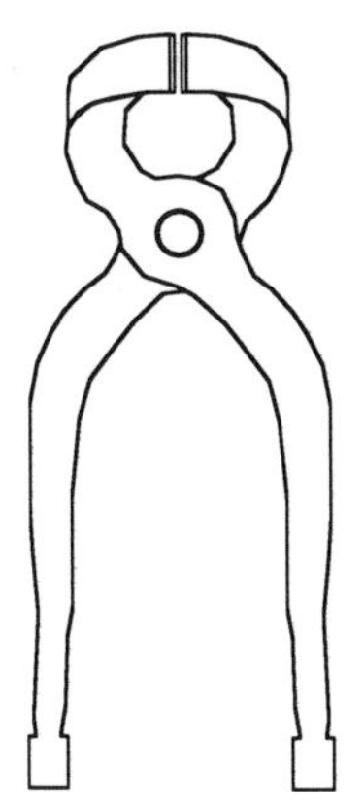

1. ____________________

2. ____________________

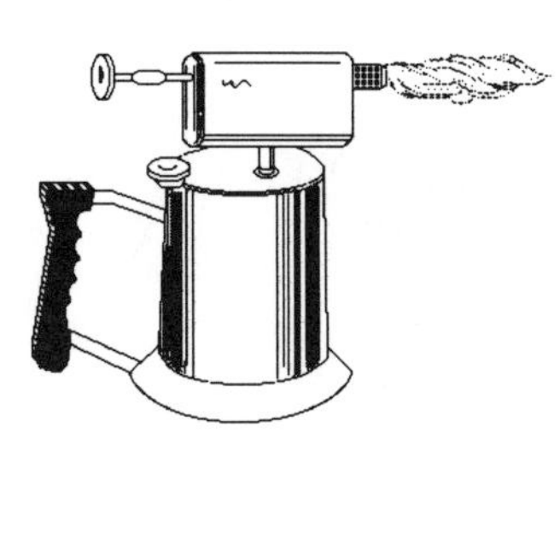

3. ____________________

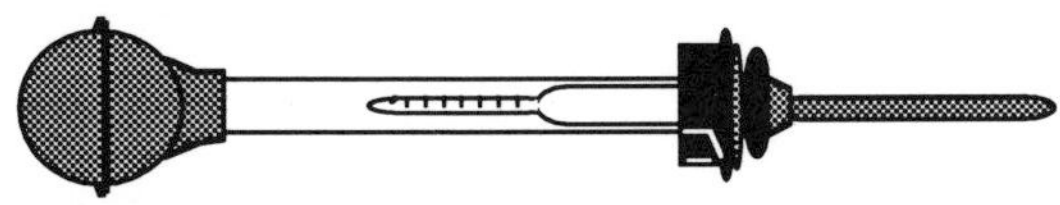

4. ____________________

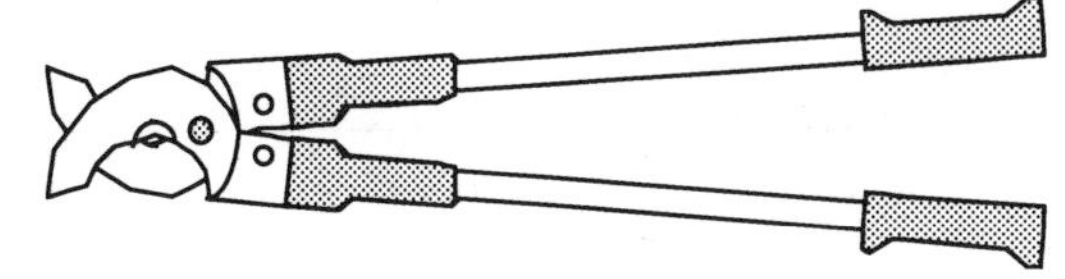

5. ____________________

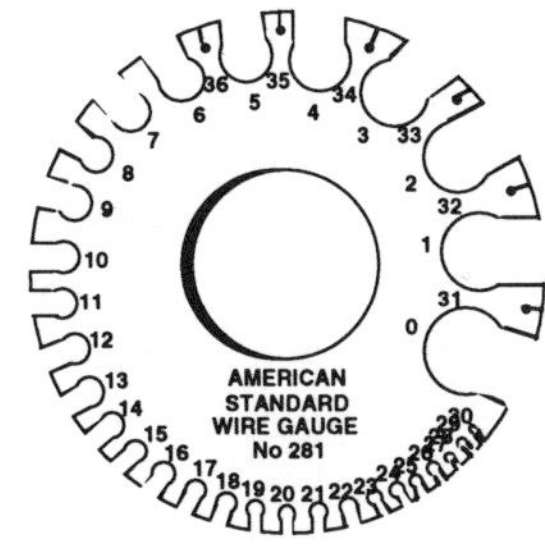

6. ____________________

7. ____________________

JOURNEYMAN WIREMAN QUIZ #16
BLUEPRINT SYMBOLS

•Fill in the blank with the correct name for the symbol

1. S_P ____________________

2. ____________________

3. (F) ____________________

4. F ____________________

5. ____________________

6. ____________________

7. ____________________

8. TV ____________________

9. ____________________

10. — — — ____________________

JOURNEYMAN WIREMAN QUIZ #17
WIRING METHODS

• *Fill in the blank with the correct name for the wiring method or equipment.*

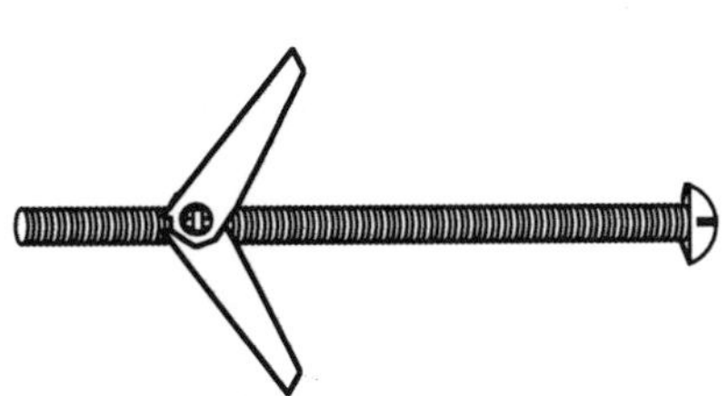

1. ______________________________

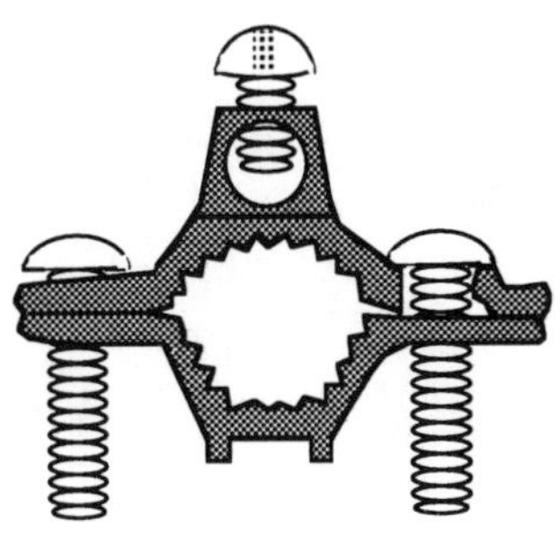

2. ______________________________

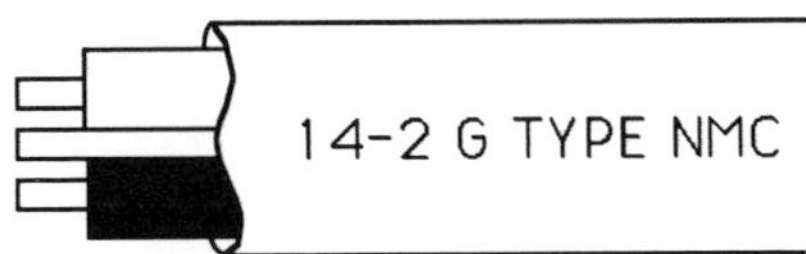

3. ______________________________

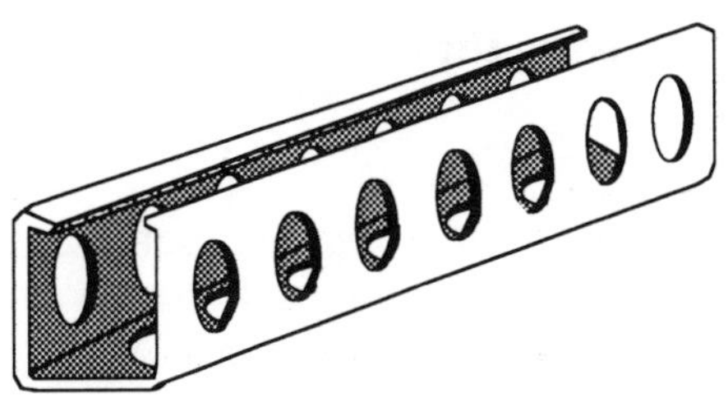

4. ______________________________

TYPE FCC UL 12 AWG, CU 300 V, 20 AMP, 60° C

CU 300 V 20 AMP 60° C AMP INC P/N 553239 TYPE FCC

300 V 20 AMP 60° C AMP INC P/N 553239

AWG, CU 300 V, 20 AMP, 60° C AMP INC P/N 553239

AMP INC P/N 553239 TYPE FCC UL 12 AWG, CU

5. ______________________________

6. ______________________________

JOURNEYMAN WIREMAN QUIZ #18
PRACTICAL QUESTIONS

•*Circle the correct answer:*

1. What is the function of a neon glow tester?

I. determines if circuit is alive
II. determines polarity of DC circuits
III. determines if circuit is AC or DC current

(a) I only (b) II only (c) III only (d) I, II and III

2. A conduit coupling is sometimes tightened by using a strap wrench rather than a Stillson wrench. The strap wrench is used when it is important to avoid ____.

(a) crushing the conduit
(b) bending the conduit
(c) stripping the threads
(d) damaging the outside finish

3. With respect to a common light bulb, it is correct to state that the ____.

(a) circuit voltage has no effect on the life of the bulb
(b) base has a left hand thread
(c) filament is made of carbon
(d) lower wattage bulb has the higher resistance

4. A multimeter is a combination of ____.

(a) ammeter, ohmmeter and wattmeter
(b) voltmeter, ohmmeter and ammeter
(c) voltmeter, ammeter and megger
(d) voltmeter, wattmeter and ammeter

5. Two 120 volt light bulbs connected in series across 240 volt will ____.

(a) burn at full brightness
(b) burn at half-brightness
(c) burn out quickly
(d) flicker with the cycle

JOURNEYMAN WIREMAN QUIZ #19
TOOL IDENTIFICATION QUESTIONS

• *Fill in the blank with the correct name for the tool.*

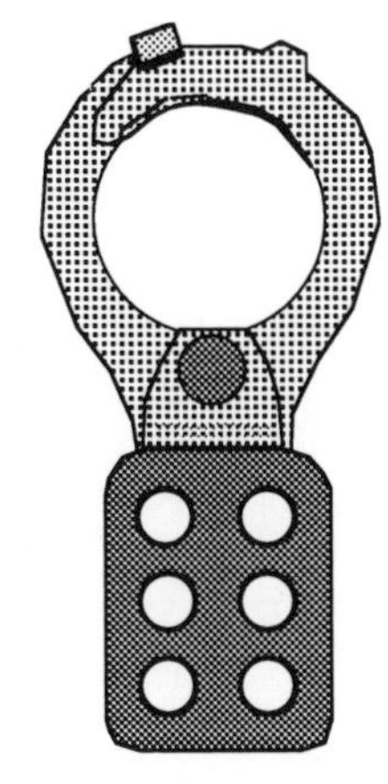

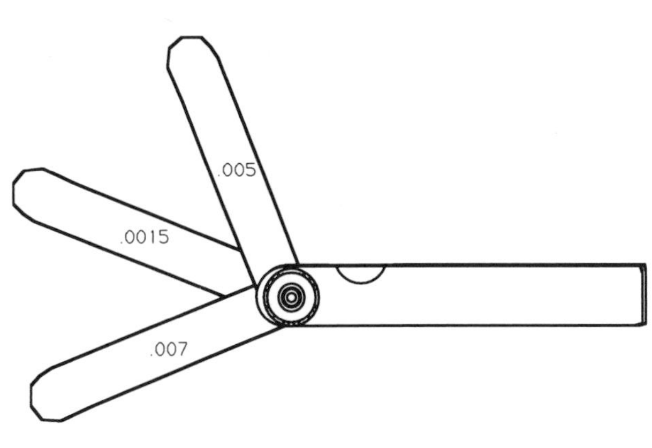

1. ______________________

2. ______________________

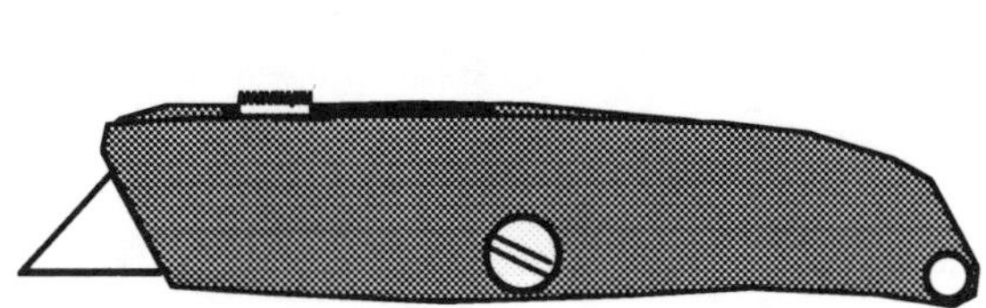

3. ______________________

4. ______________________

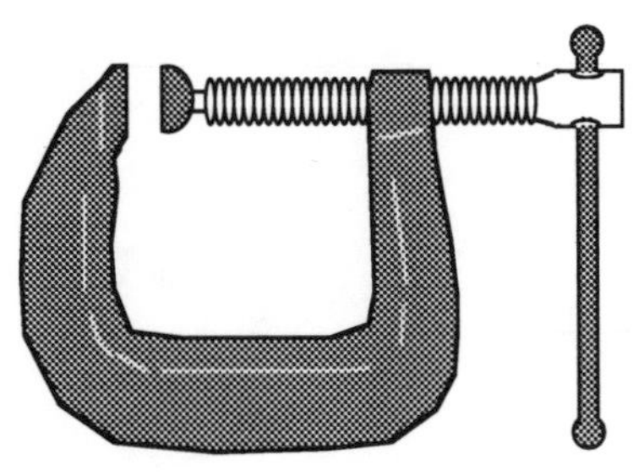

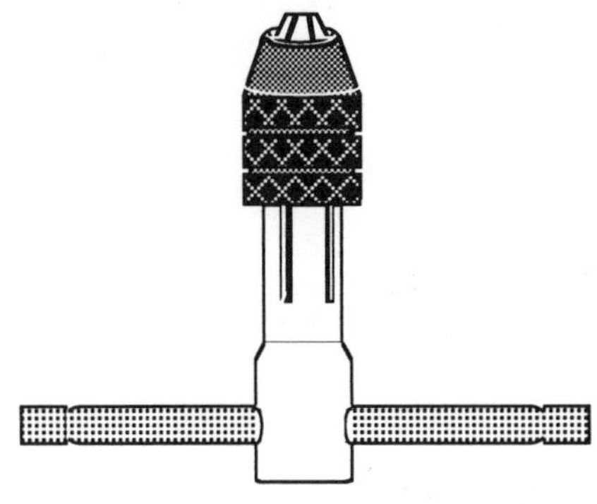

5. ______________________

6. ______________________

JOURNEYMAN WIREMAN QUIZ #20
BLUEPRINT SYMBOLS

•*Fill in the blank with the correct name for the symbol*

1. (B) ____

2. ____

3. ____

4. ____

5. S_T ____

6. CH ____

7. ____

8. ____

9. ____

10. 3 ____

JOURNEYMAN WIREMAN QUIZ #21
CONTROLS

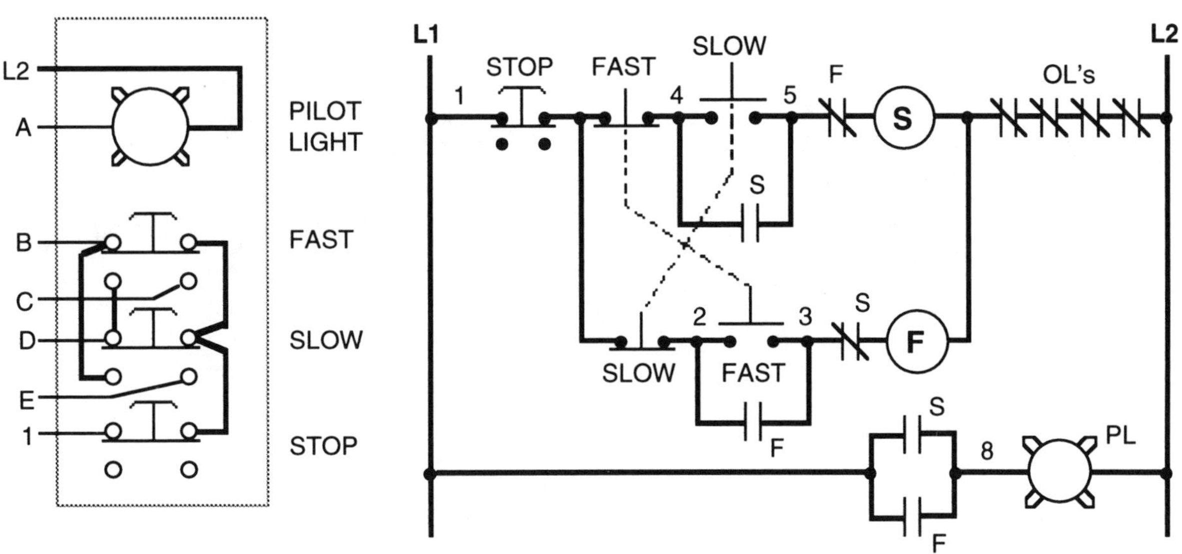

1. "**A**" is actually point _____ on the control diagram.
   a) L1 b) L2 c) 8 d) 6

2. "**D**" is actually point _____ on the control diagram.
   a) 5 b) 4 c) 3 d) 2

3. "**B**" is actually point _____ on the control diagram.
   a) 4 b) 3 c) 2 d) L1

4. "**C**" is actually point _____ on the control diagram.
   a) 4 b) 3 c) 2 d) L1

5. "**E**" is actually point _____ on the control diagram.
   a) 4 b) 5 c) 1 d) 2

JOURNEYMAN WIREMAN QUIZ #22
TOOL IDENTIFICATION QUESTIONS

• *Fill in the blank with the correct name for the tool.*

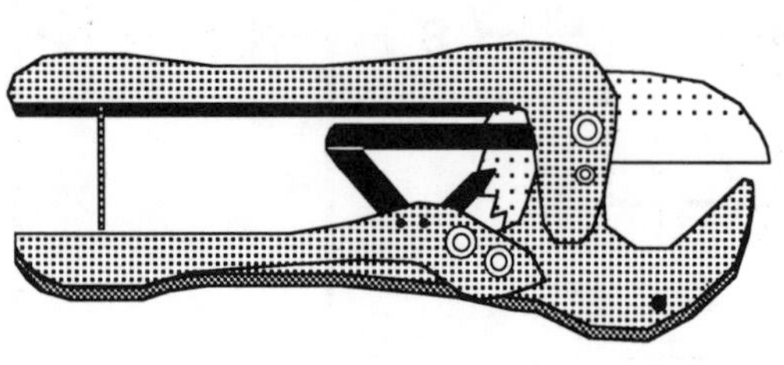

1. ______________________________

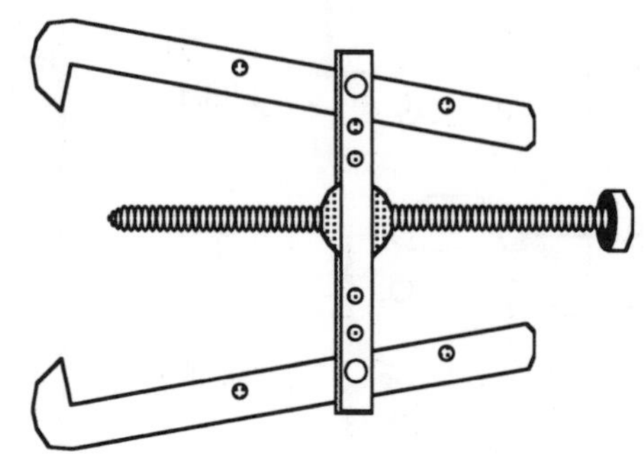

2. 

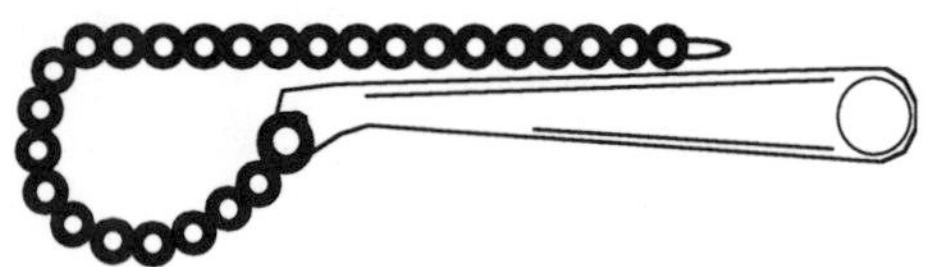

3. ______________________________

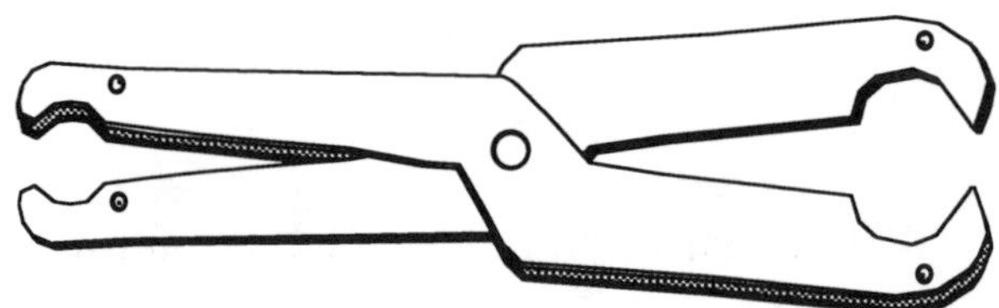

4. 

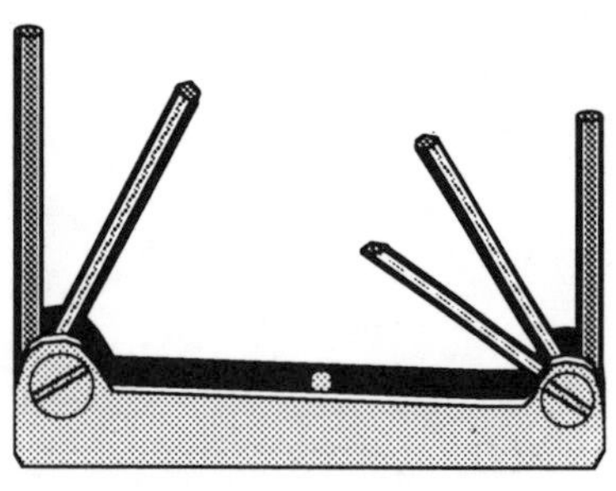

5. ______________________________

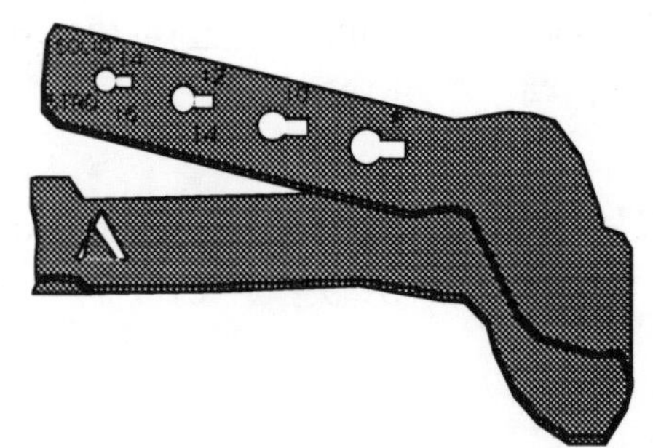

6. ______________________________

JOURNEYMAN WIREMAN QUIZ #23
PRACTICAL QUESTIONS

•*Circle the correct answer:*

1. What type fastener would you use to mount a box to a hollow tile wall?

(a) expansion bolts
(b) wooden plugs
(c) toggle bolts
(d) bolts with backing plates

2. When a current leaves its intended path and returns to the source bypassing the load, the circuit is ____.

(a) open
(b) shorted
(c) incomplete
(d) broken

3. A clamp-on ammeter will measure ____.

(a) voltage when clamped on a single conductor
(b) current when clamped on a multiconductor cable
(c) accurately only when parallel to the cable
(d) accurately only when clamped perpendicular to a conductor

4. Which of the following statements about mounting single-throw knife switches in a vertical position is/are correct?

I. the switch shall be mounted so that the blade hinge is at the bottom
II. the supply side of the circuit shall be connected to the bottom of the switch

(a) I only (b) II only (c) both I and II (d) neither I nor II

5. A hook on the end of a fish tape is **not** to ____.

(a) keep it from catching on joints and bends
(b) tie a swab to
(c) tie the wires, to be pulled, to
(d) protect the end of the wire

JOURNEYMAN WIREMAN QUIZ #24
BLUEPRINT SYMBOLS

•Fill in the blank with the correct name for the symbol

1. ________________________________

2. ________________________________

3. ________________________________

4.  GR ________________________________

5. ________________________________

6.  C ________________________________

7. ________________________________

8. ________________________________

9. ________________________________

10.  PS ________________________________

JOURNEYMAN WIREMAN QUIZ #25
TOOL IDENTIFICATION QUESTIONS

• *Fill in the blank with the correct name for the tool.*

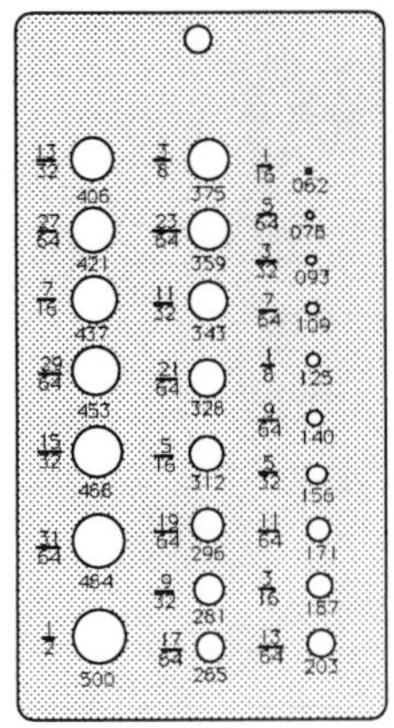

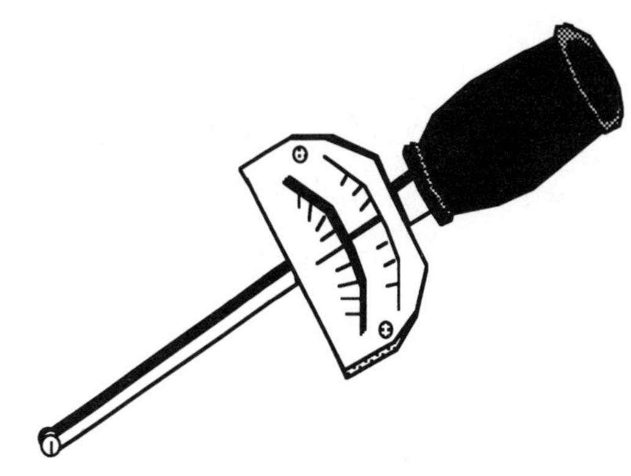

1. ______________________________

2. ______________________________

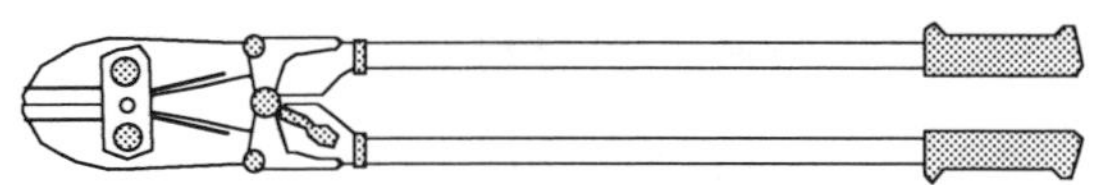

3. ______________________________

4. ______________________________

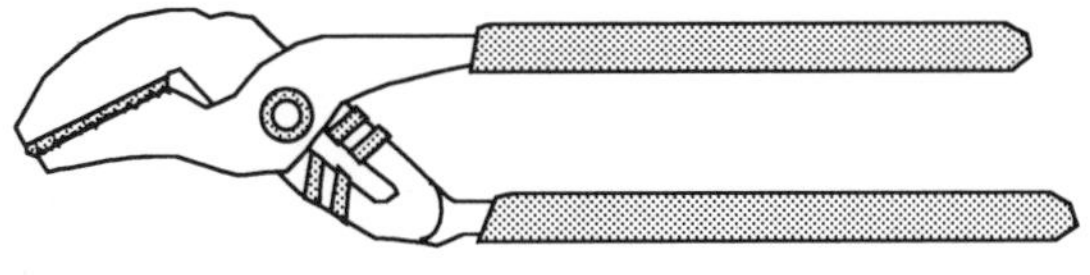

5. ______________________________

6. ______________________________

JOURNEYMAN WIREMAN QUIZ #26
WIRING METHODS

• *Fill in the blank with the correct name for the wiring method or equipment.*

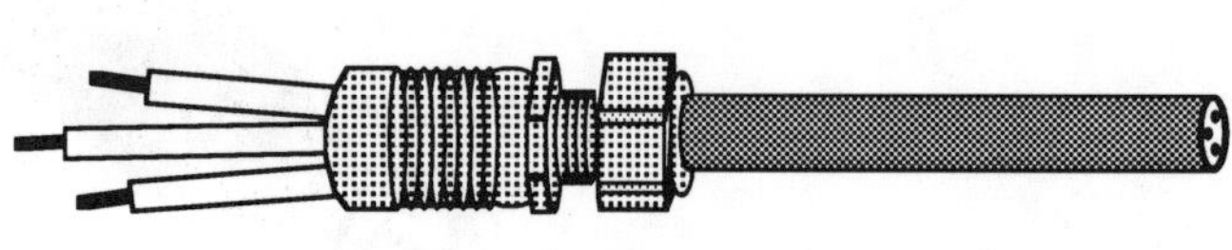

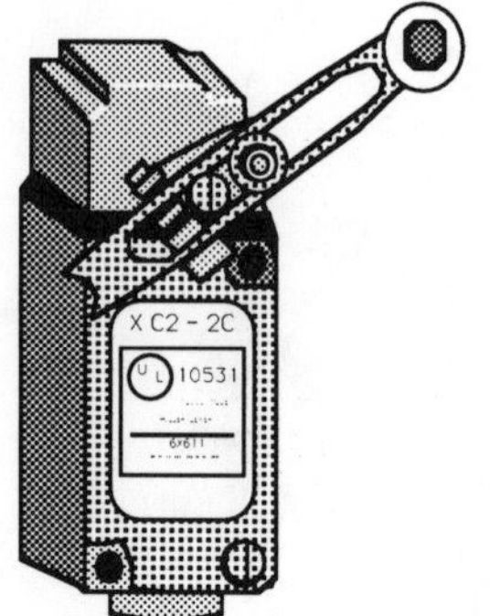

1. ______________________________

2. ______________________________

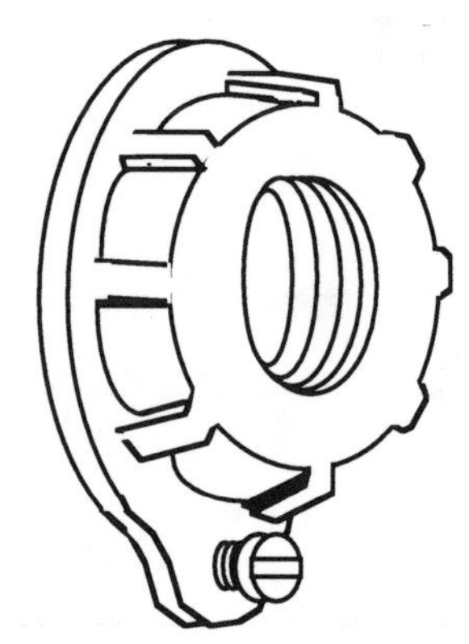

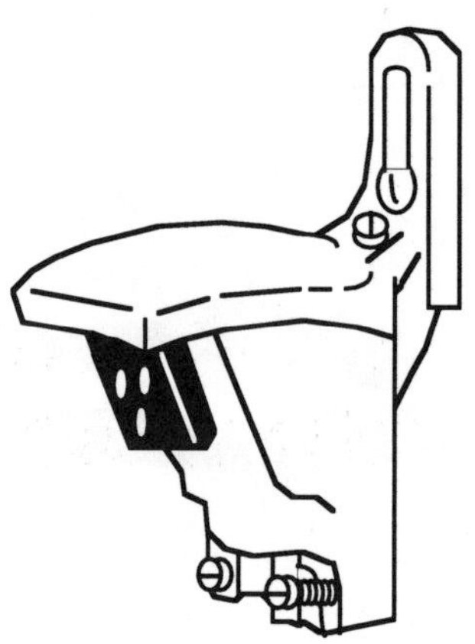

3. ______________________________

4. ______________________________

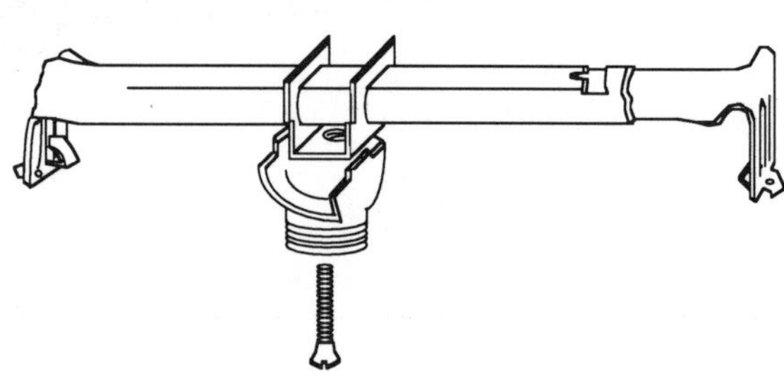

5. ______________________________

6. ______________________________

JOURNEYMAN WIREMAN QUIZ #27

• *The top watthour meter is read at the beginning of the month, the bottom meter is read at the end of the month. How many kilowatthours were consumed?* ________ *kwh.*

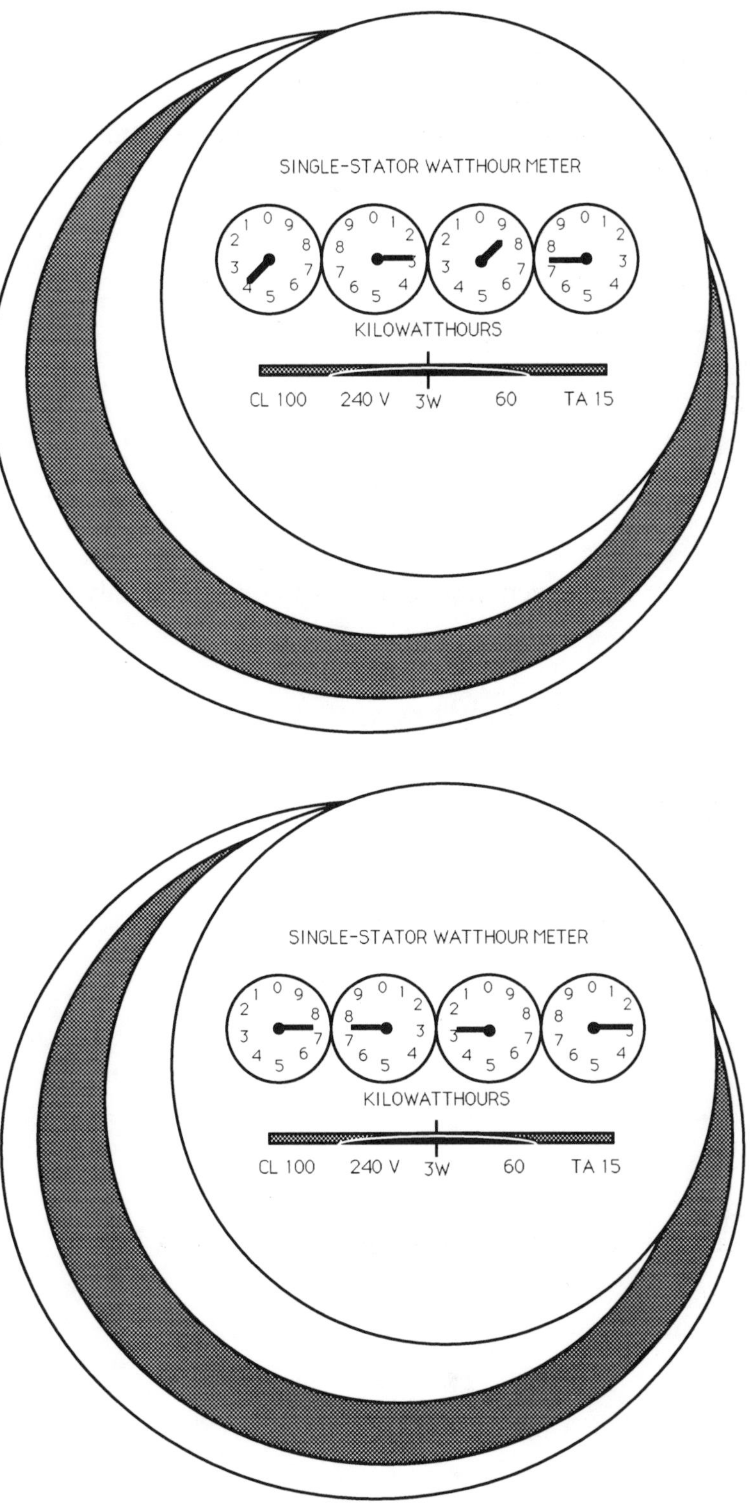

JOURNEYMAN WIREMAN QUIZ #28
PRACTICAL QUESTIONS

•*Circle the correct answer:*

1. An Erickson coupling is used to ____.

(a) join sections of EMT together
(b) connect EMT to flexible conduit
(c) to connect two sections of rigid conduit when one section cannot be turned
(d) substitute for all-thread

2. On smaller gauges of wire, they are pencil-stripped to prevent ____.

(a) hysteresis
(b) over stripping
(c) nicks in wire
(d) loosening of wire nut

3. When the term "10-32" in connection with machine bolts commonly used in lighting work, the number "32" refers to ____.

(a) bolt length
(b) bolt thickness
(c) diameter of hole
(d) threads per inch

4. A pendant light fixture is a _____ fixture.

(a) closet
(b) recessed
(c) hanging
(d) bracket

5. Since fuses are rated by amperage and voltage a fuse will work on ____.

(a) AC only
(b) AC or DC
(c) DC only
(d) any voltage

JOURNEYMAN WIREMAN QUIZ #29
CONTROLS

•*Fill in the blank below for the correct type of motor starting.*

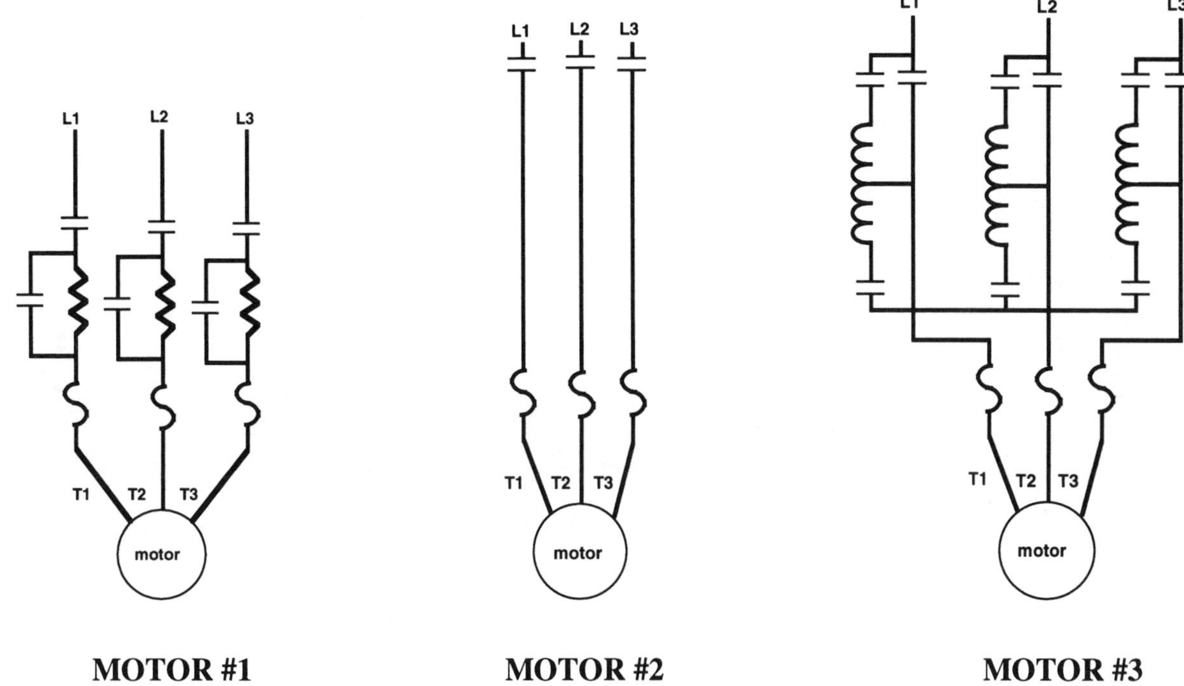

MOTOR #1 **MOTOR #2** **MOTOR #3**

1. Motor #1 is a _______________ type start.

(a) full voltage (b) autotransformer (c) resistor

2. Motor #2 is a _______________ type start.

(a) full voltage (b) autotransformer (c) resistor

3. Motor #3 is a _______________ type start.

(a) full voltage (b) autotransformer (c) resistor

JOURNEYMAN WIREMAN QUIZ #30
TOOL IDENTIFICATION QUESTIONS

• *Fill in the blank with the correct name for the tool.*

1. 

2. 

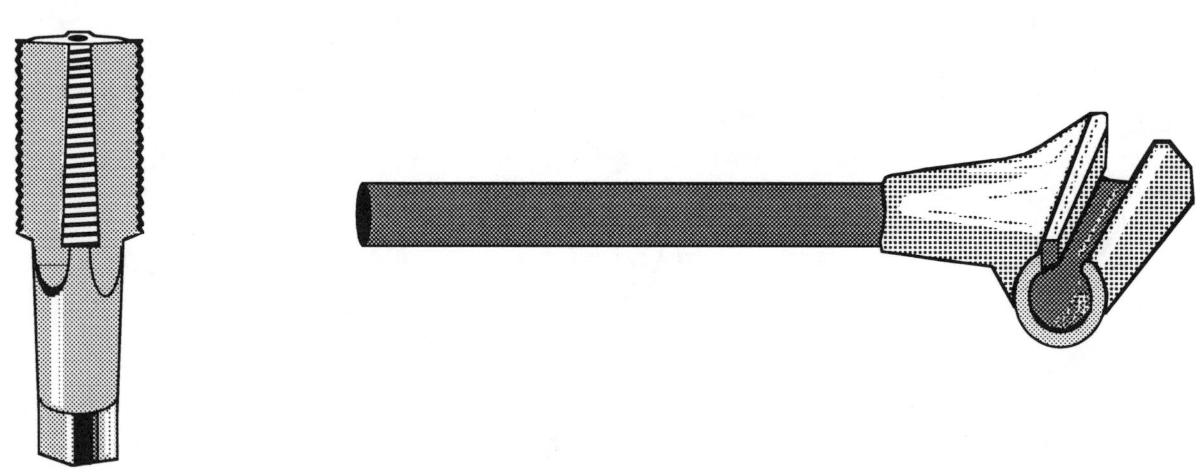

3. ______________________________

4. ______________________________

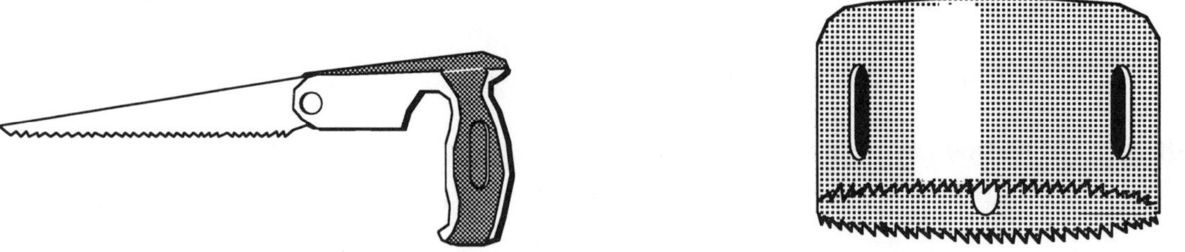

5. ______________________________

6. ______________________________

JOURNEYMAN WIREMAN QUIZ #31
INSTALLATION

•Circle the correct installation method.

Which of the following is the correct practice to tighten a nut with an adjustable wrench?

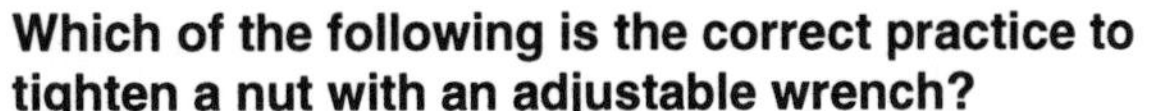

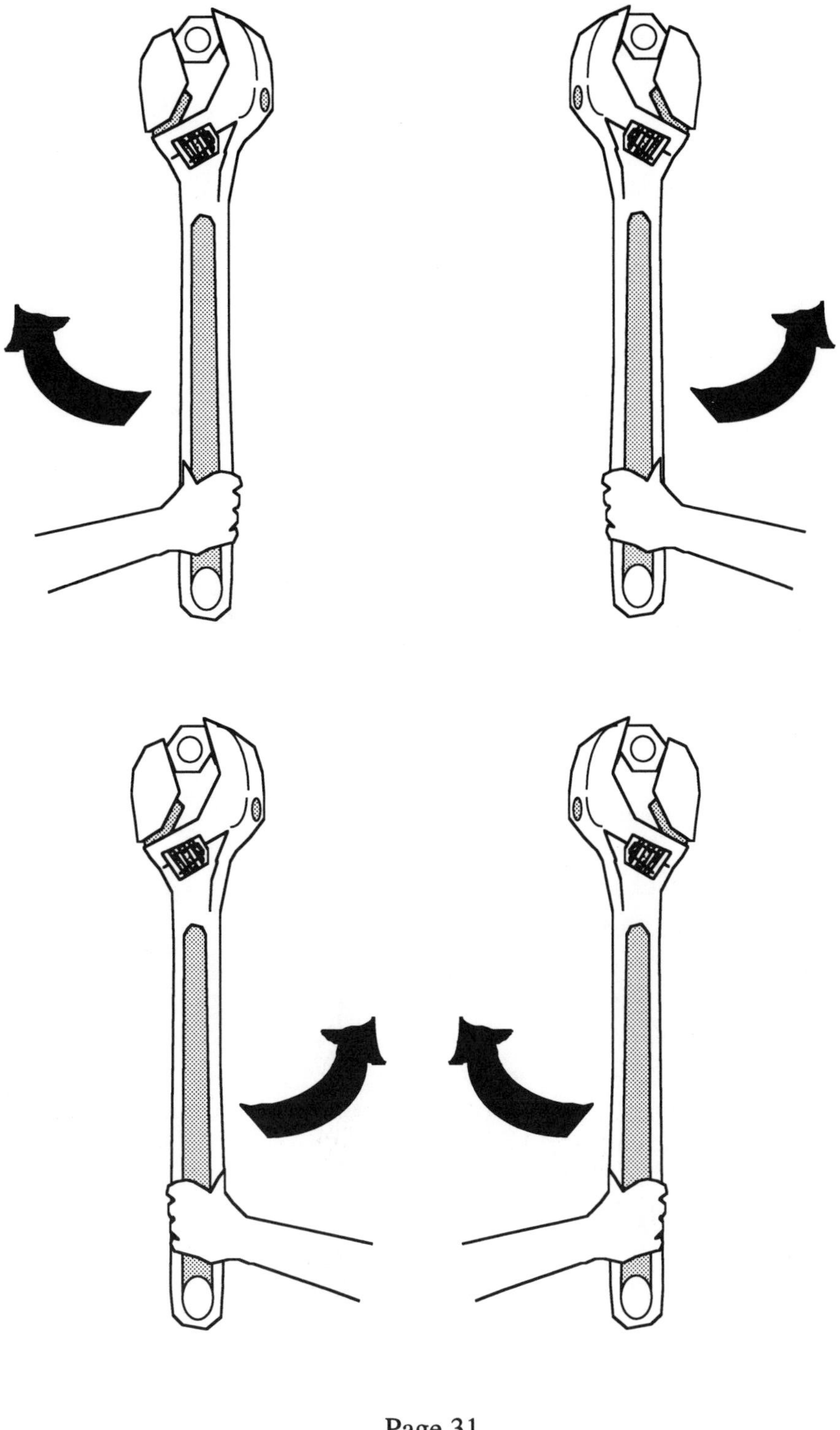

JOURNEYMAN WIREMAN QUIZ #32
PRACTICAL QUESTIONS

•*Circle the correct answer:*

1. The neutral conductor shall not be ____.

(a) stranded
(b) solid
(c) insulated
(d) fused

2. The part of an electrical system that performs a mechanical function rather than an electrical function is called a(n) ____.

(a) receptacle
(b) device
(c) fitting
(d) outlet

3. Solid wire is preferred instead of stranded wire in panel wiring because ____.

(a) costs less than stranded
(b) solid will carry more current
(c) can be "shaped" better
(d) no derating required for solid

4. What is meant by "traveler wires"?

(a) wiring to a split receptacle
(b) two-wires between 3-way switches
(c) wiring to a door bell
(d) out of state electrician

5. When working near acid storage batteries, extreme care should be taken to guard against sparks, essentially to avoid ____.

(a) overheating the electrolyte
(b) an electric shock
(c) a short circuit
(d) an explosion

JOURNEYMAN WIREMAN QUIZ #33
TESTING

Which of the fuses is blown?

- *Circle the line that the fuse is* *<u>BLOWN.</u>*   *L1*   *or*   *L2*

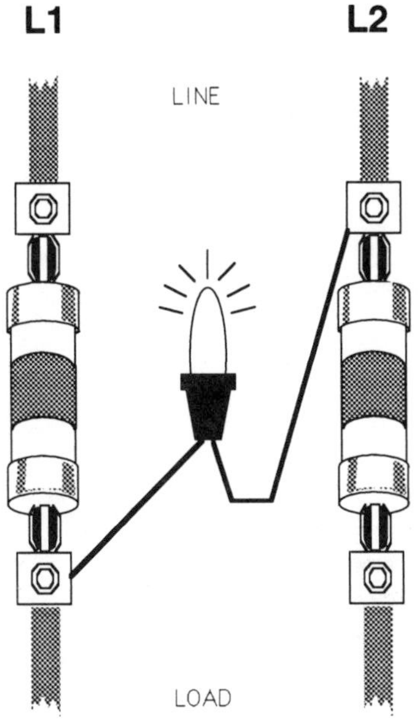

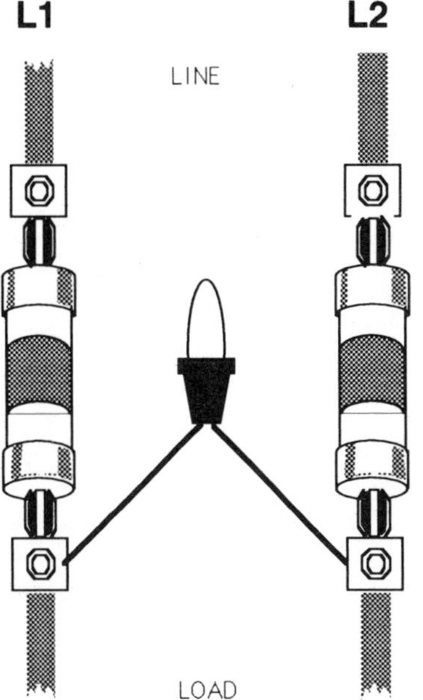

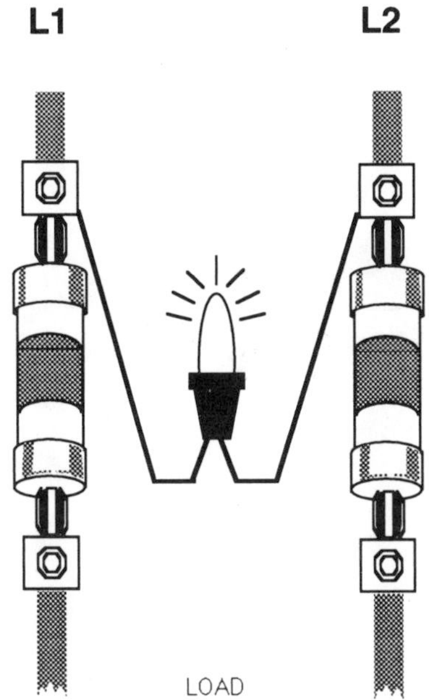

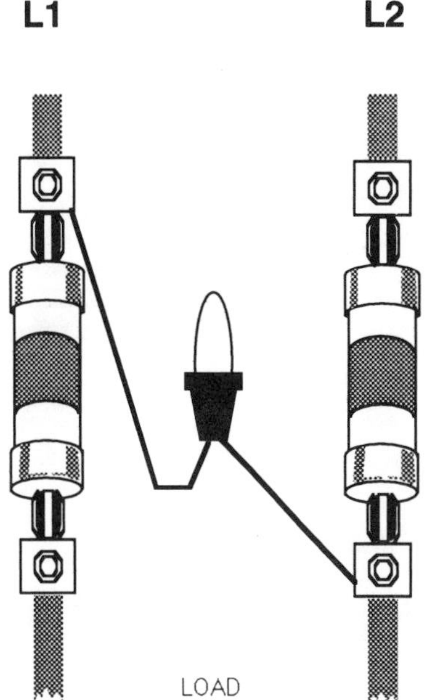

JOURNEYMAN WIREMAN QUIZ #34
INSTALLATION

•Circle the correct installation method.

Which of the following is the correct practice to splice a cord?

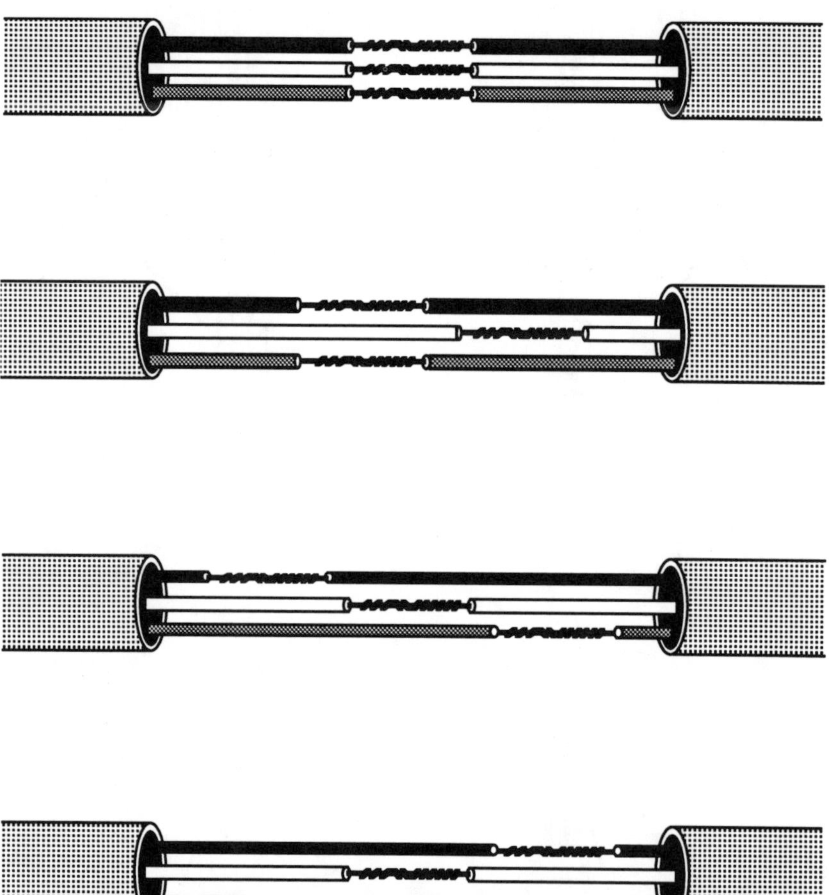

JOURNEYMAN WIREMAN QUIZ #35
TOOL IDENTIFICATION QUESTIONS

• *Fill in the blank with the correct name for the tool.*

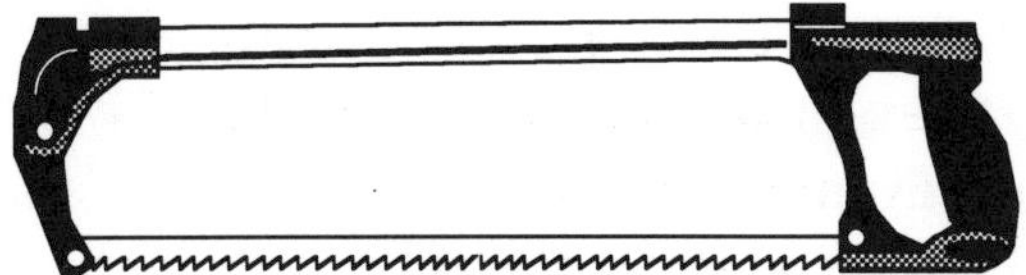

1. ______________________________

2. ______________________________

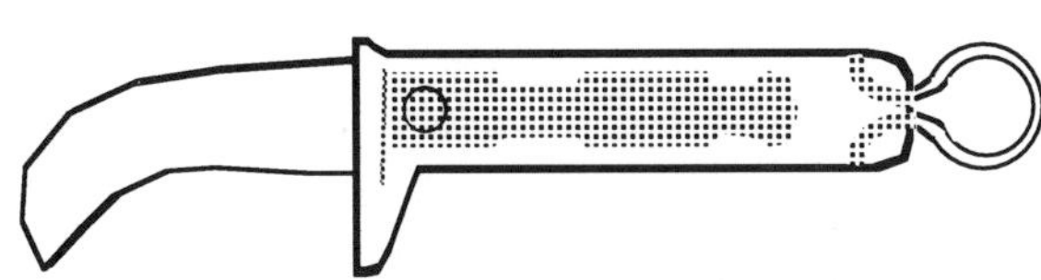

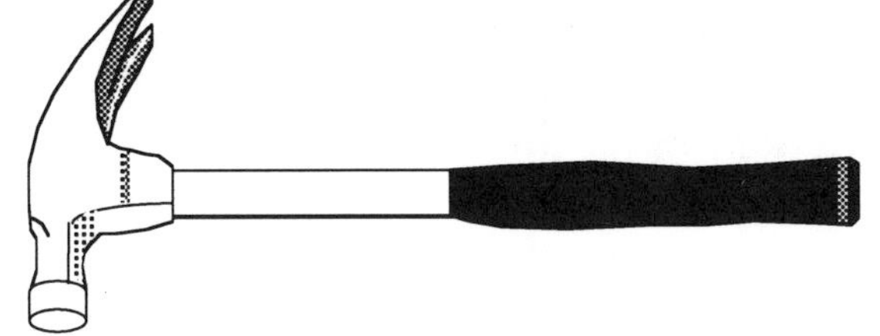

3. ______________________________

4. ______________________________

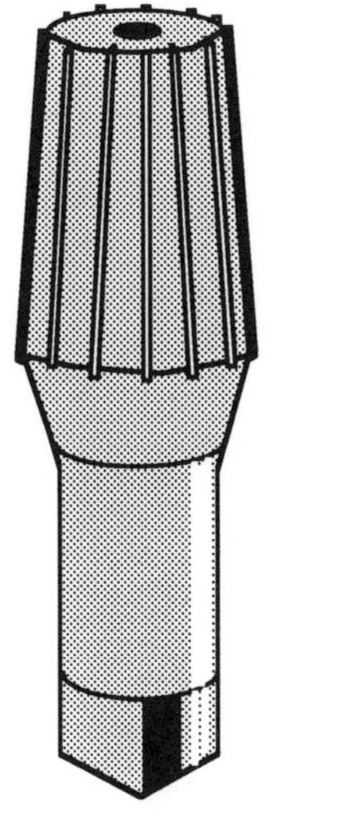

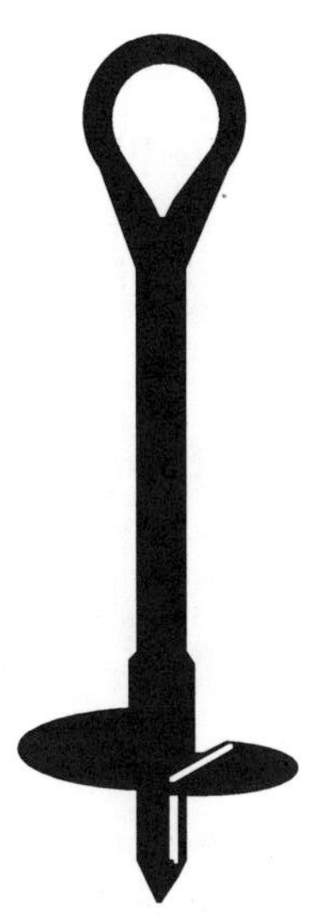

5. ______________________________

6. ______________________________

JOURNEYMAN WIREMAN QUIZ #36
PRACTICAL QUESTIONS

•*Circle the correct answer:*

1. The reason for grounding the frame of a portable electric hand tool is to ____.

(a) prevent the frame of the tool from becoming alive to ground
(b) prevent overheating of the tool
(c) prevent shorts
(d) reduce voltage drop

2. The purpose of a Western Union splice is ____.

(a) for the use of the utility companies only
(b) for the purpose of strengthening a splice
(c) for the use on the west coast only
(d) none of the above

3. To mark a point on the floor directly beneath a point on the ceiling, it is best to use a ____.

(a) transit rod
(b) plumb bob
(c) square
(d) 12' tape

4. When installing an instrument meter on a panel, to obtain accurate mounting ____.

(a) use the meter and drill thru the holes
(b) drill oversize holes
(c) use a template
(d) drill from back of panel

5. The advantage of cutting a metal rigid conduit with a hacksaw rather than a pipe cutter is ____.

(a) you do not need a vice
(b) less energy required in cutting
(c) less reaming is required
(d) threading oil is not required

JOURNEYMAN WIREMAN QUIZ #37
CONTROL SYMBOLS

•Fill in the blank with the correct letter from choices below for the symbol

1. ______ 2. ______ 3. ______ 4. ______ 5. ______ 6. ______

7. ______ 8. ______ 9. ______ 10. ______ 11. ______ 12. ______

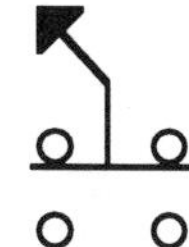

13. ______ 14. ______ 15. ______ 16. ______ 17. ______ 18. ______

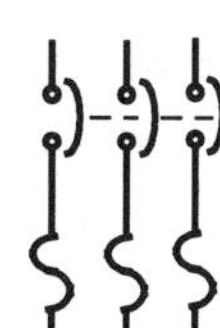

19. ______ 20. ______

•Choose a letter ( ) and fill in the blank above:

(A) CB with thermal O.L.
(B) normally closed contact
(C) liquid level switch N.C.
(D) temperature actuated switch N.O.
(E) foot switch N.C.
(F) start button N.O.
(G) Timed contact N.O.T.C.
(H) stop button N.C.
(I) disconnect
(J) limit switch N.O.
(K) SPDT double break
(L) normally open contact
(M) temperature actuated switch N.C.
(N) thermal O.L.
(O) selector switch-two position
(P) foot switch N.O.
(Q) limit switch N.C.
(R) autotransformer winding
(S) liquid level switch N.O.
(T) mushroom head push button switch

JOURNEYMAN WIREMAN QUIZ #38
TOOL IDENTIFICATION QUESTIONS

• *Fill in the blank with the correct name for the tool.*

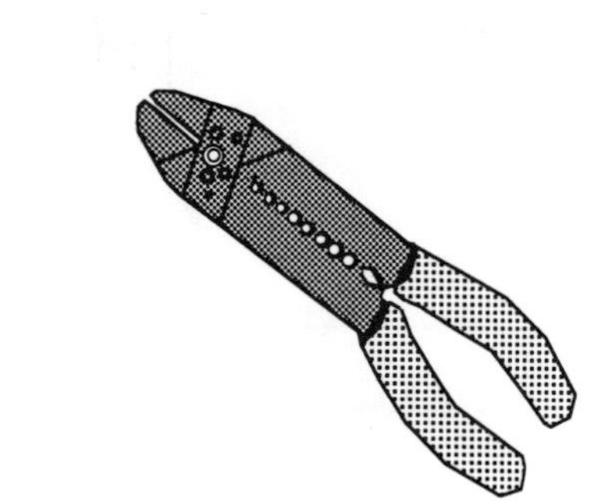

1. ______________________________

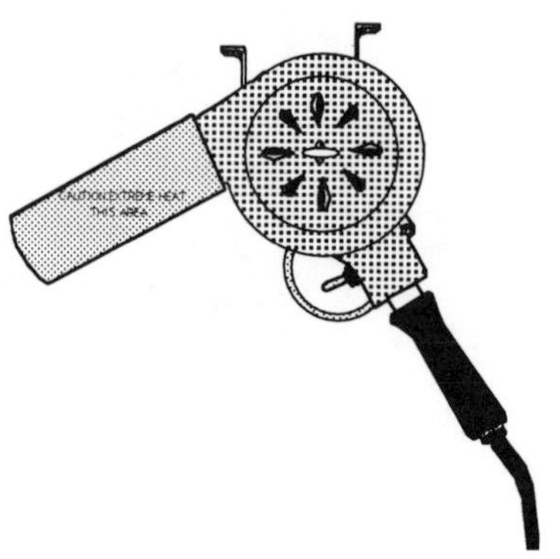

2. ______________________________

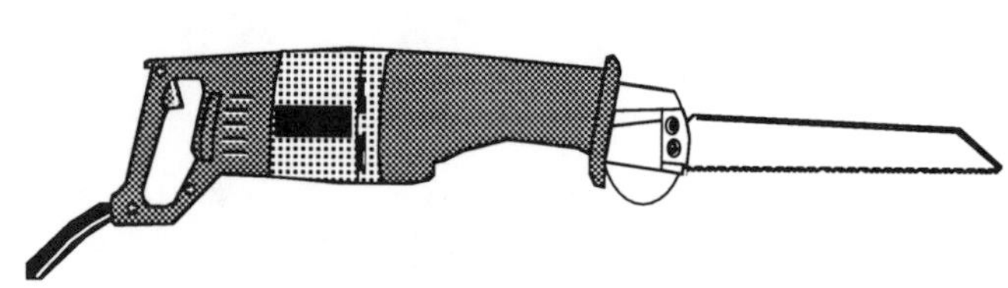

3. ______________________________

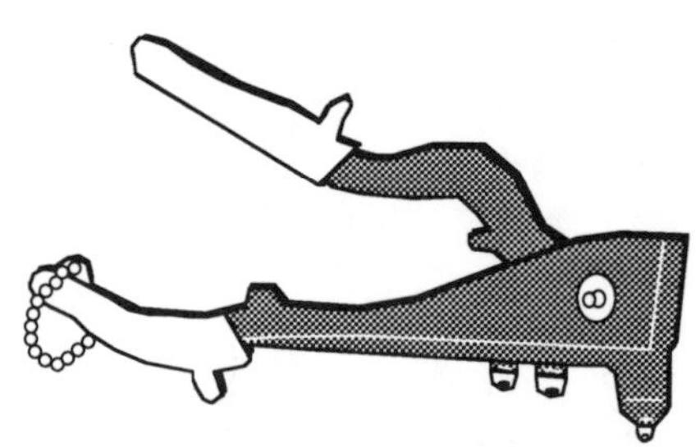

4. ______________________________

5. ______________________________

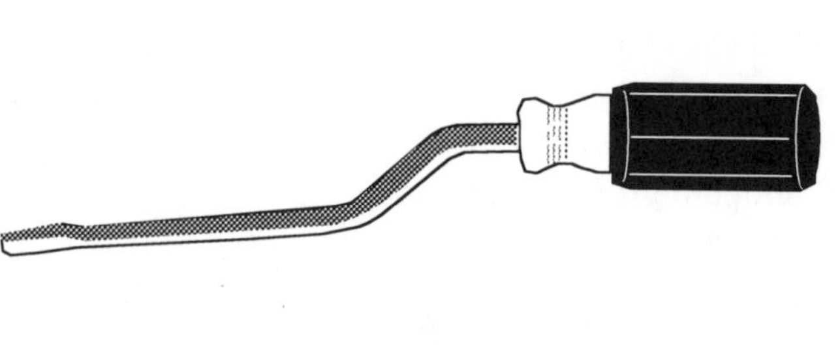

6. ______________________________

JOURNEYMAN WIREMAN QUIZ #39
INSTALLATION

•Circle the correct installation method.

Which of the following is the correct practice to loosen a pipe with a pipe wrench ?

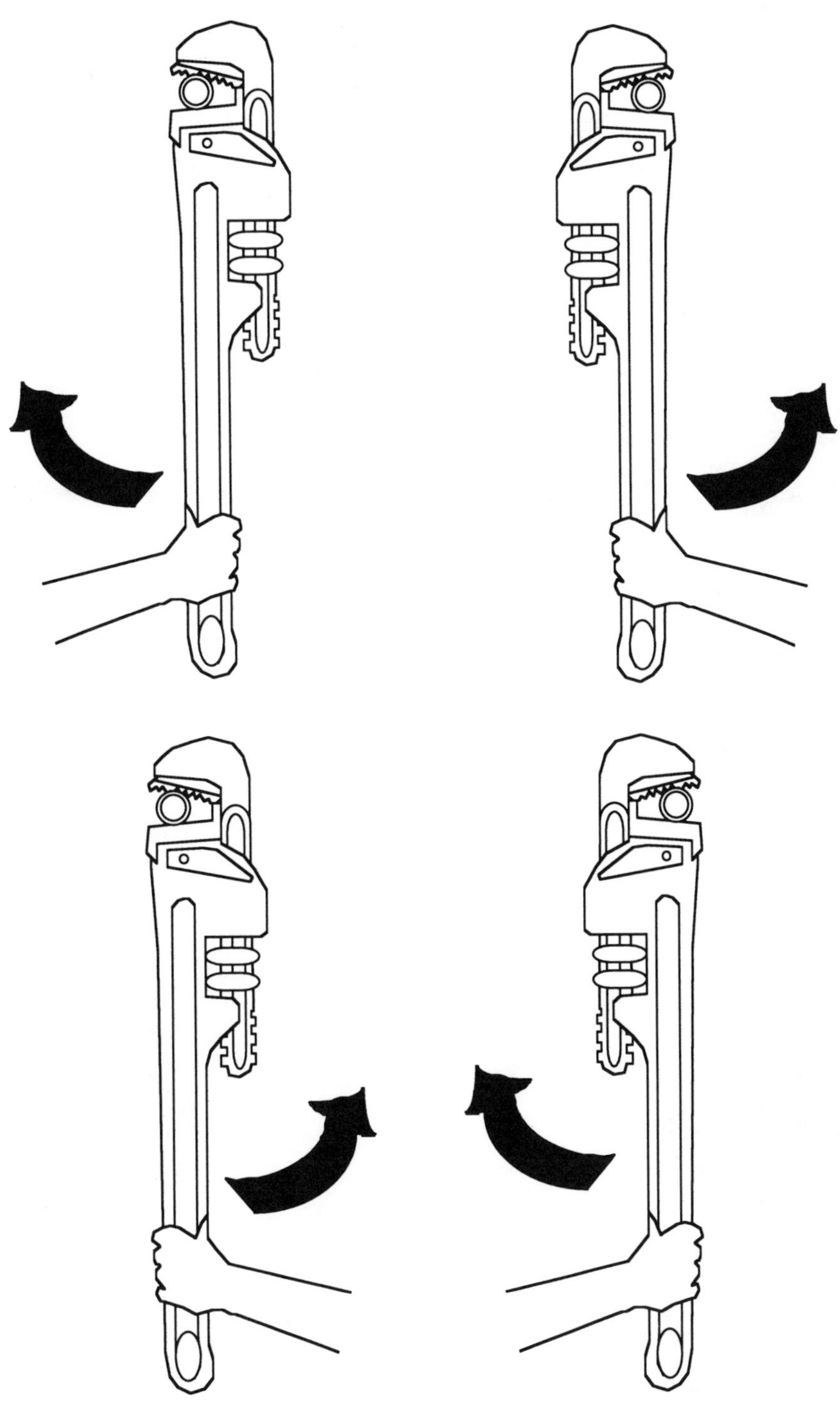

JOURNEYMAN WIREMAN QUIZ #40
TOOL IDENTIFICATION QUESTIONS

• *Fill in the blank with the correct name for the tool.*

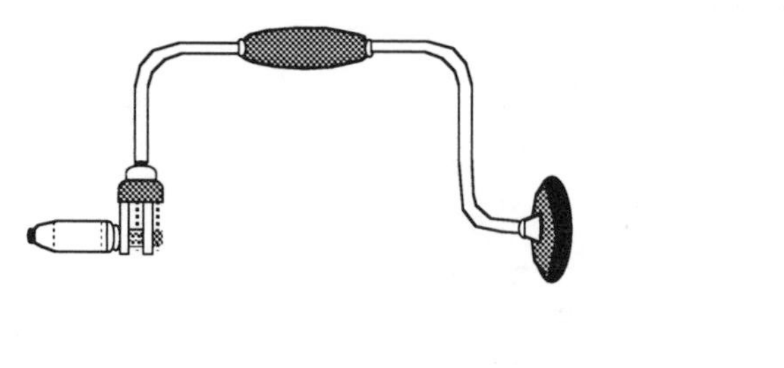

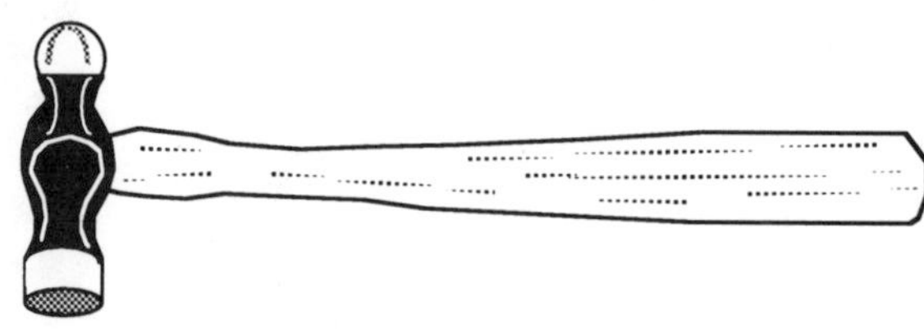

1. ______________________________

2. 

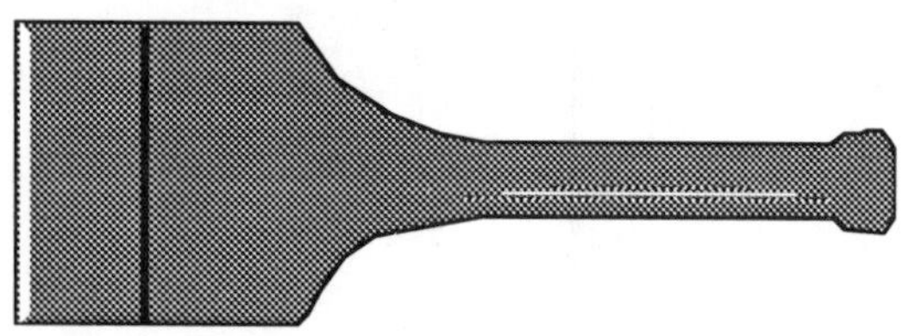

3. ______________________________

4. 

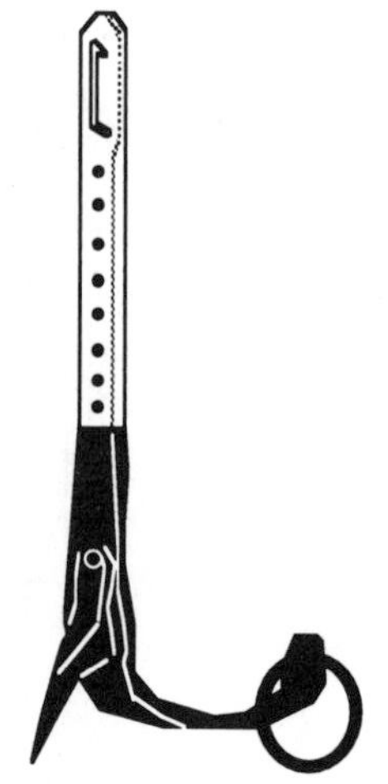

5. ______________________________

6. ______________________________

JOURNEYMAN WIREMAN QUIZ #41
PRACTICAL QUESTIONS

•*Circle the correct answer:*

1. Multiple start buttons in a motor control circuit are connected in ____.

(a) series
(b) parallel
(c) series-parallel
(d) none of the above

2. A function of a relay is to ____.

(a) turn on another circuit
(b) produce thermal electricity
(c) limit the flow of electrons
(d) create a resistance in the field winding

3. To control a ceiling light from five different locations it requires which of the following?

(a) four 3-way switches and one 4-way switch
(b) three 4-way switches and two 3-way switches
(c) three 3-way and two 4-way switches
(d) four 4-way switches and one 3-way switch

4. The identified grounded conductor of a lighting circuit is always connected to the screw of a light socket to ____.

(a) reduce the possibility of accidental shock
(b) ground the light fixture
(c) improve the efficiency of the lamp
(d) provide the easiest place to connect the wire

5. If the end of a cartridge fuse becomes warmer than normal, you should ____.

(a) tighten the fuse clips
(b) lower the voltage on the circuit
(c) notify the ultility company
(d) change the fuse

JOURNEYMAN WIREMAN QUIZ #42
WIRING METHODS

• *Fill in the blank with the correct letter for the wiring method or equipment:*

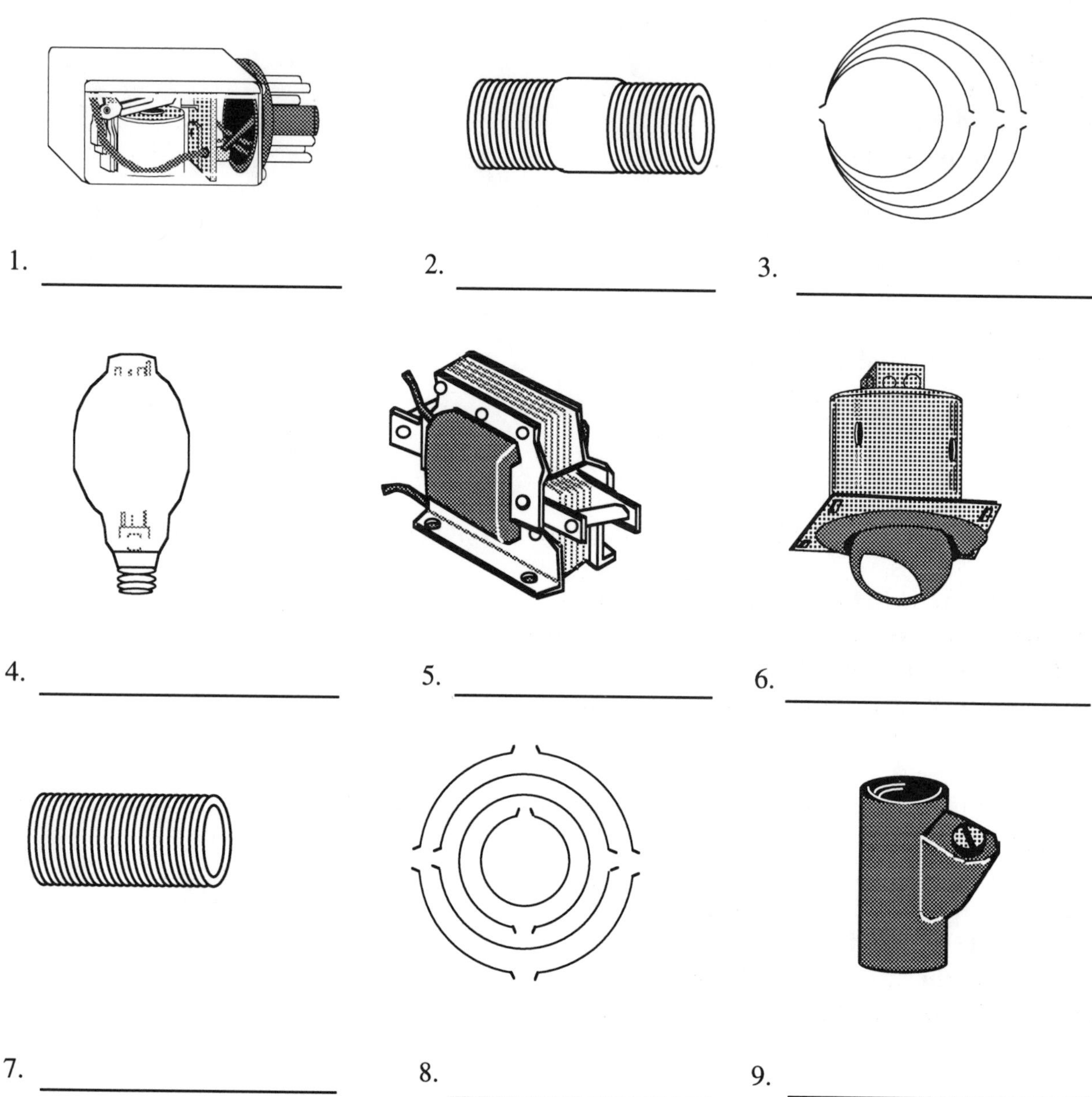

• *Choose a letter ( ) and fill in the blank above:*

(A) concentric K.O.
(B) close nipple
(C) mercury lamp
(D) eccentric K.O.
(E) relay
(F) seal off fitting
(G) short nipple
(H) solenoid
(I) bull eye's light

JOURNEYMAN WIREMAN QUIZ #43
TOOL IDENTIFICATION QUESTIONS

• *Fill in the blank with the correct name for the tool.*

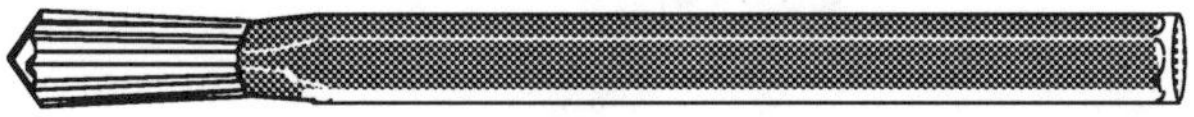

1. ______________________________

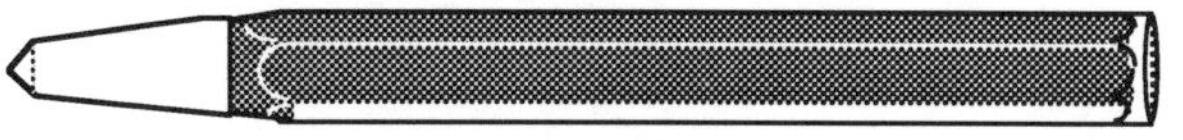

2. ______________________________

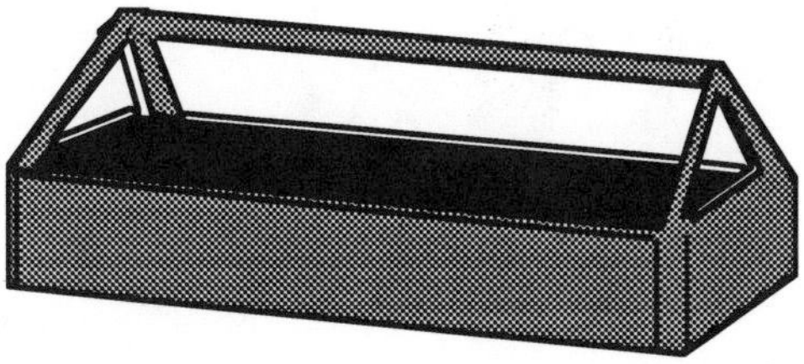

3. ______________________________

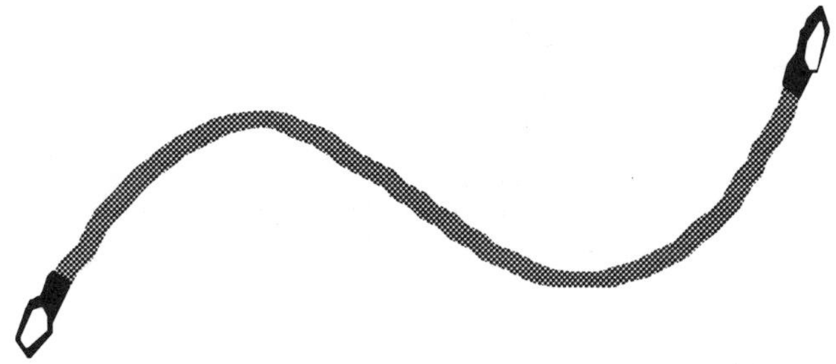

4. ______________________________

5. ______________________________

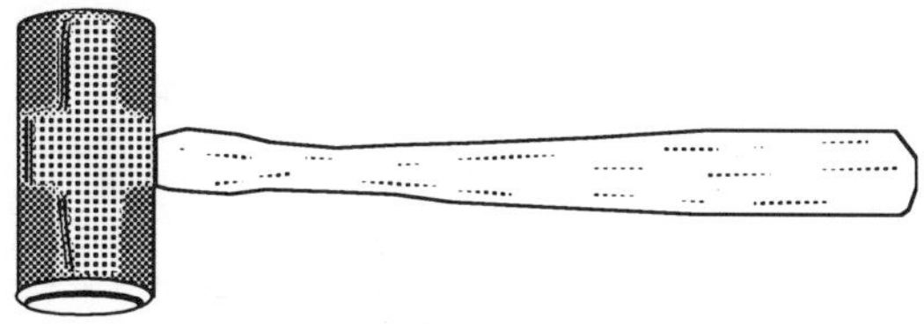

6. ______________________________

JOURNEYMAN WIREMAN QUIZ #44
INSTALLATION

•*Circle the correct installation method.*

Which of the following is the correct practice when cutting thin wall conduit?

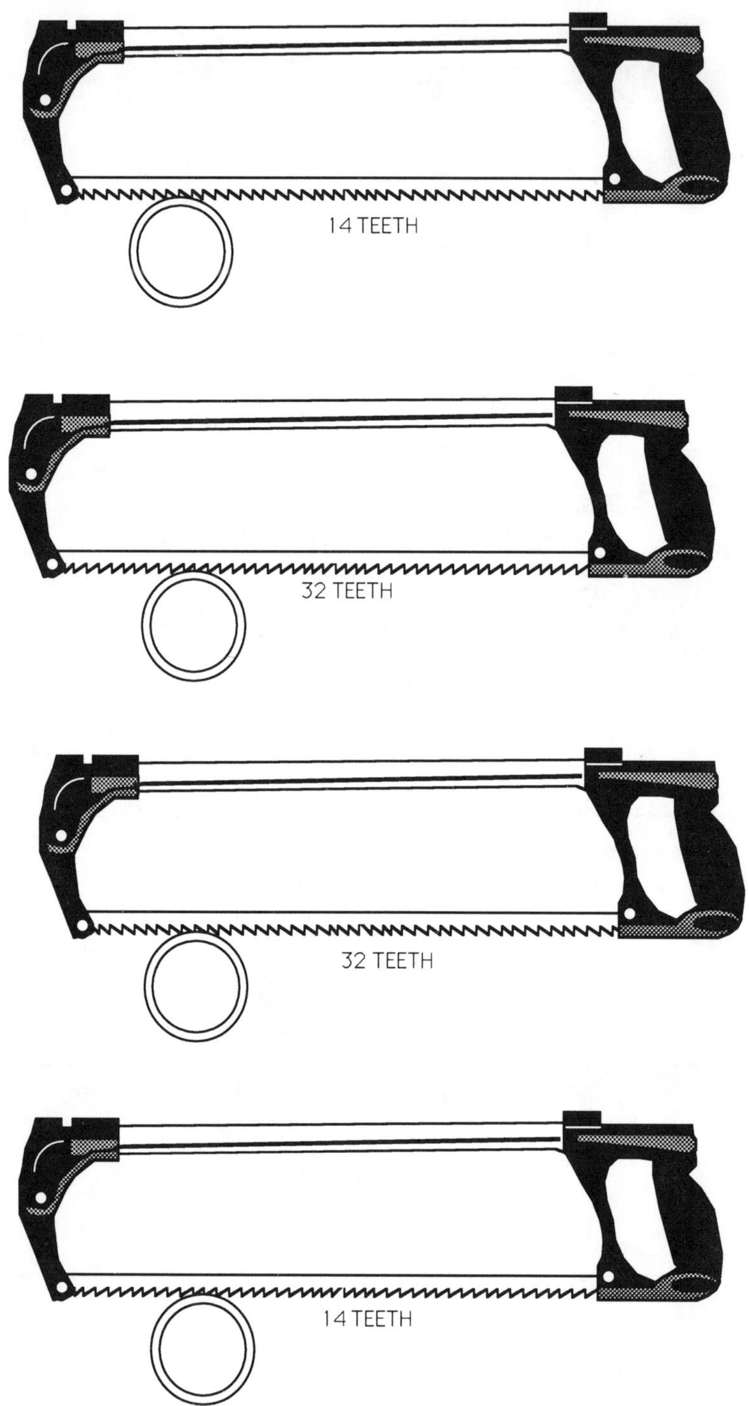

JOURNEYMAN WIREMAN QUIZ #45
PRACTICAL QUESTIONS

•*Circle the correct answer:*

1. The reason for installing electrical conductors in a conduit is _____.

(a) to provide a ground
(b) to increase the ampacity of the conductors
(c) to protect the conductors from damage
(d) to avoid derating for continuous loading of conductors

2. A ladder which is painted is a safety hazard mainly because the paint ____.

(a) may conceal weak spots in the rails or rungs
(b) is slippery after drying
(c) causes the wood to crack more quickly
(d) peels and the sharp edges of the paint may cut the hands

3. A conduit body is ____.

(a) a cast fitting such as an FD or FS box
(b) a standard 10 foot length of conduit
(c) a sealtight enclosure
(d) an "LB" or "T", or similar fitting

4. Raceways shall be provided with ____ to compensate for thermal expansion and contraction.

(a) accordion joints
(b) thermal fittings
(c) expansion joints
(d) contro-spansion

5. A type of cable protected by a spiral metal cover is called ____ in the field.

(a) BX
(b) greenfield
(c) sealtight
(d) Romex

JOURNEYMAN WIREMAN QUIZ #46
WIRING METHODS

• *Fill in the blank with the correct letter for the wiring method or equipment:*

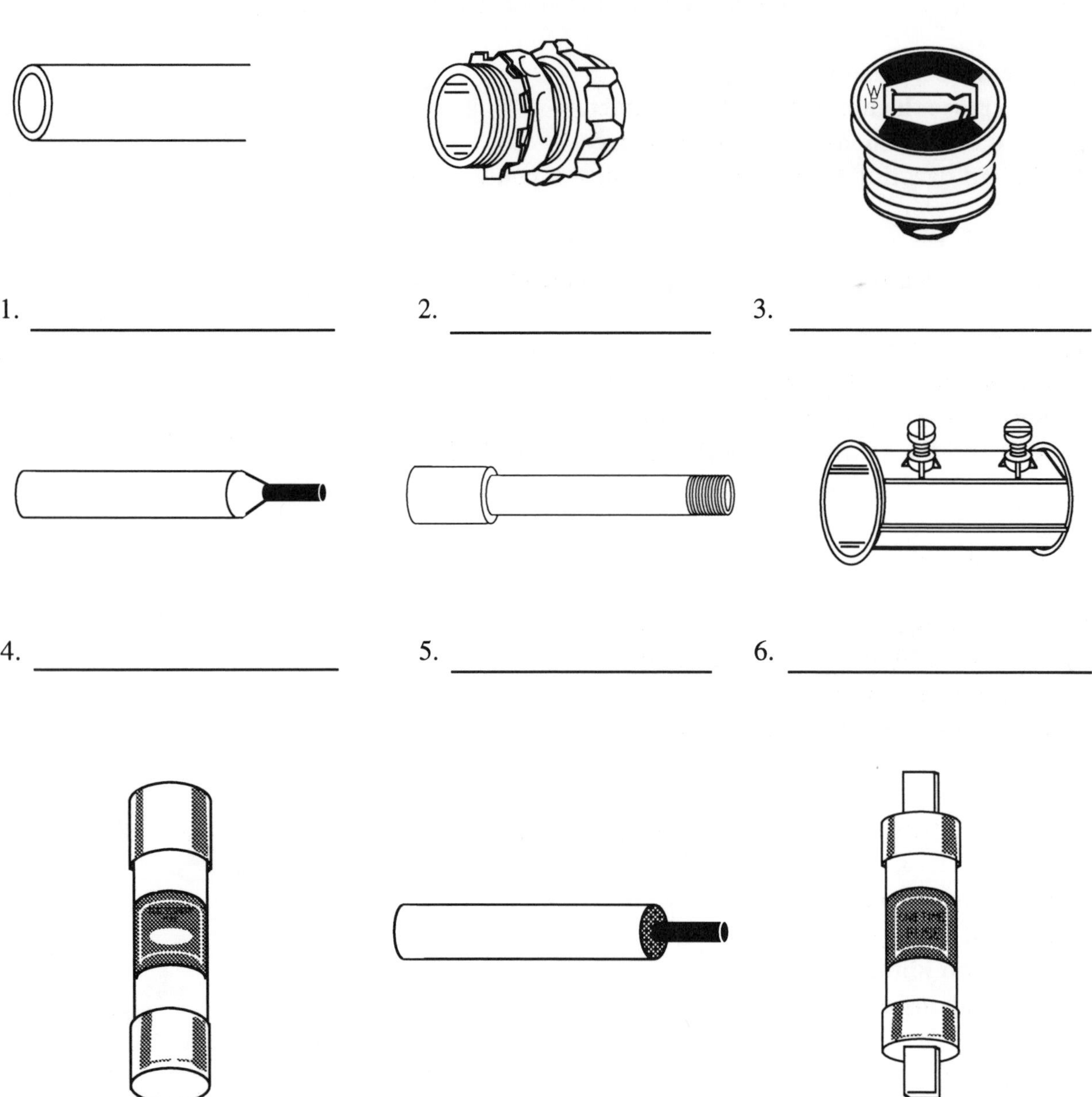

1. ______________ 2. ______________ 3. ______________

4. ______________ 5. ______________ 6. ______________

7. ______________ 8. ______________ 9. ______________

• *Choose a letter ( ) and fill in the blank above:*

(A) knife-blade fuse
(B) wrong way to skin a wire
(C) thin wall tubing
(D) coupling
(E) plug fuse
(F) right way to skin a wire
(G) ferrule-contact fuse
(H) rigid conduit
(I) box connector

JOURNEYMAN WIREMAN QUIZ #47
TOOL IDENTIFICATION QUESTIONS

• *Fill in the blank with the correct name for the tool.*

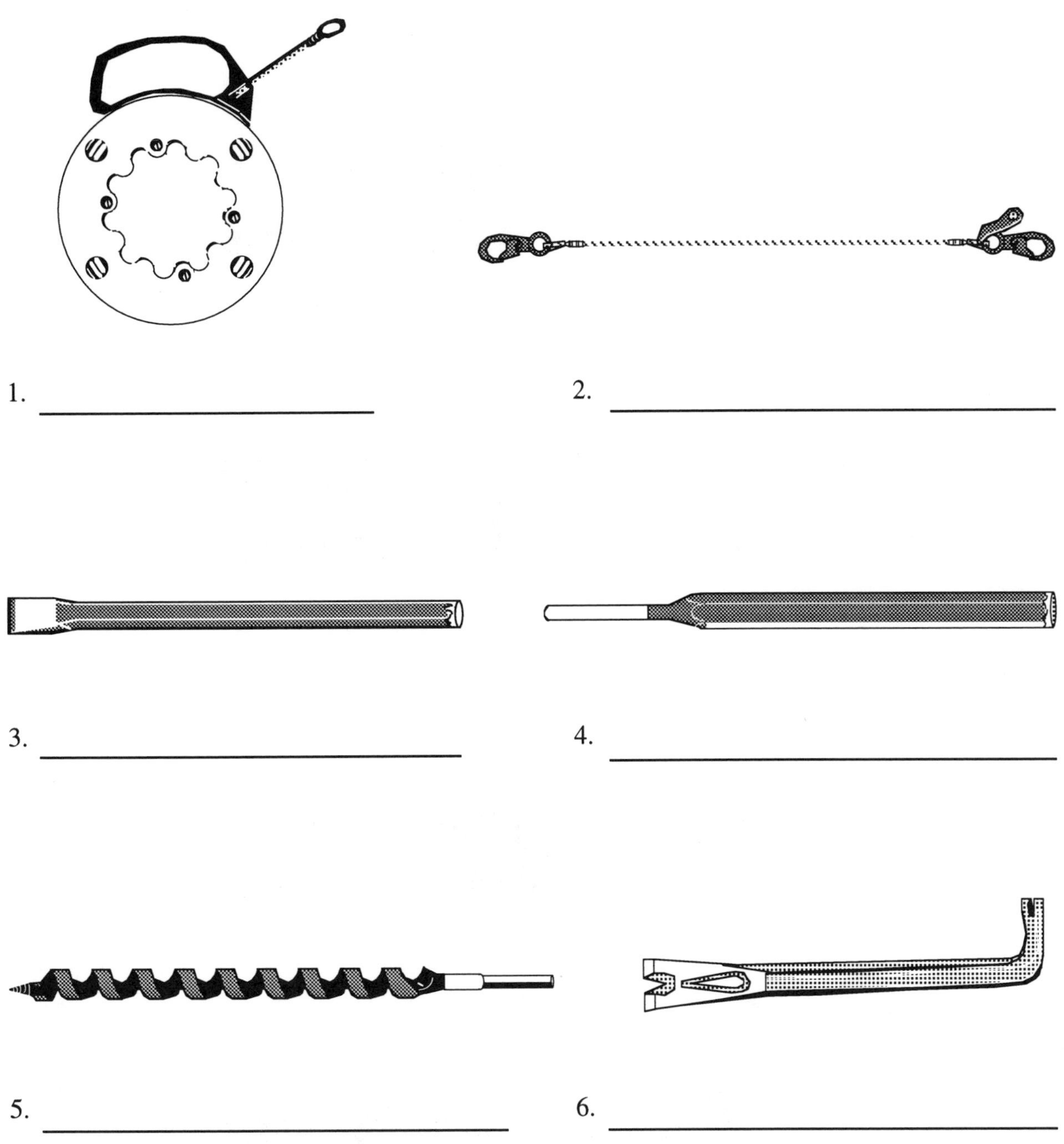

1. ______________________
2. ______________________
3. ______________________
4. ______________________
5. ______________________
6. ______________________

JOURNEYMAN WIREMAN QUIZ #48
INSTALLATION

•Circle the correct installation method.

Which of the following is the correct practice to crush the insulation prior to skinning the wire?

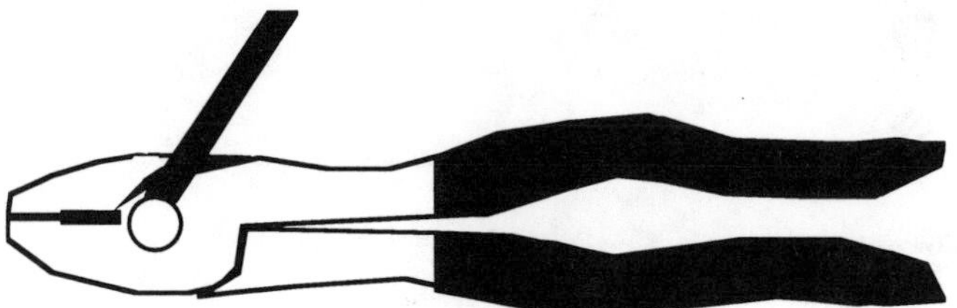

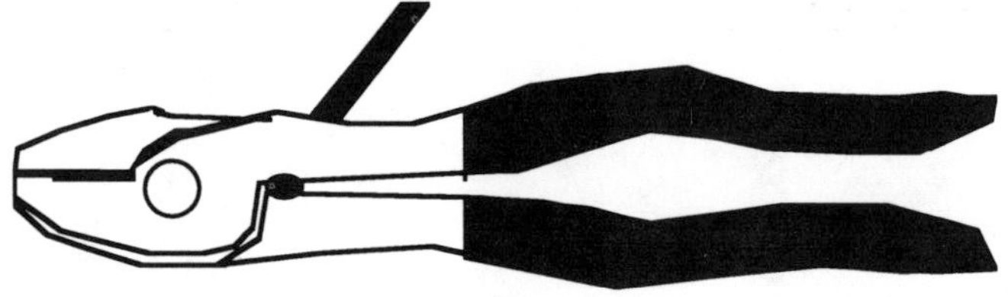

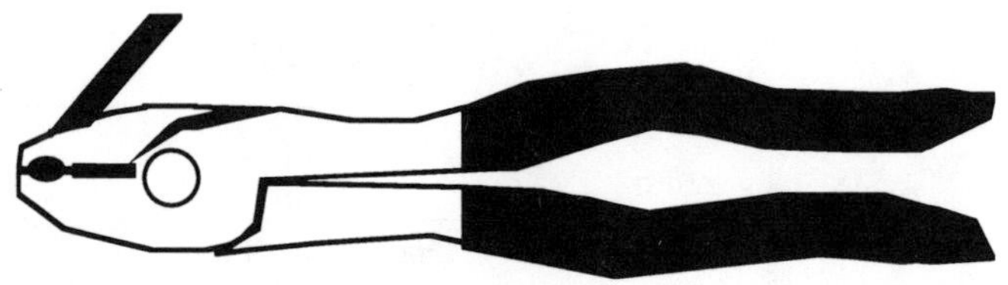

JOURNEYMAN WIREMAN QUIZ #49
PRACTICAL QUESTIONS

•*Circle the correct answer:*

1. With respect to fluorescent lamps it is correct to state ____.

(a) the filaments seldom burn out
(b) the starters and tubes must be replaced at the same time
(c) they are easier to install than incandescent light bulbs
(d) their efficiency is less than the efficiency of incandescent light bulbs

2. To increase the life of an incandescent light bulb you could ____.

(a) use at a higher than rated voltage
(b) use at a lower than rated voltage
(c) turn off when not in use
(d) use at a higher wattage

3. Which of the following hacksaw blades should be used for the best results in cutting EMT?

(a) 12 teeth per inch
(b) 18 teeth per inch
(c) 24 teeth per inch
(d) 32 teeth per inch

4. The letters DPDT are used to identify a type of ____.

(a) insulation
(b) fuse
(c) motor
(d) switch

5. When cutting a metal conduit with a hacksaw, the pressure applied to the hacksaw should be on ____.

(a) the return stroke
(b) the forward stroke only
(c) both the forward and return stroke equally
(d) none of the above

JOURNEYMAN WIREMAN QUIZ #50
TOOL IDENTIFICATION QUESTIONS

• *Fill in the blank with the correct name for the tool.*

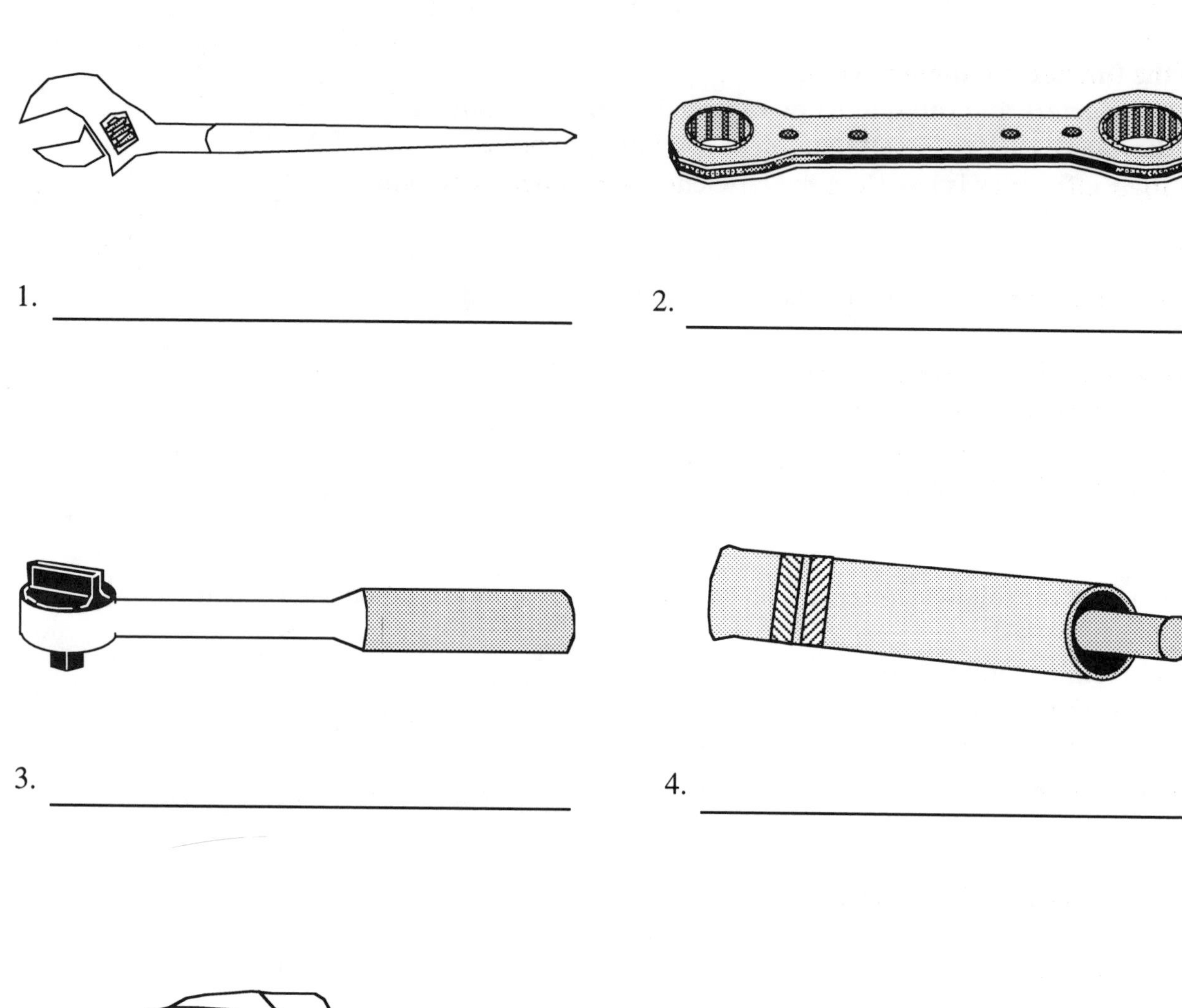

1. ______________________________

2. ______________________________

3. ______________________________

4. ______________________________

5. ______________________________

6. ______________________________

JOURNEYMAN WIREMAN QUIZ #51
INSTALLATION

•Circle the correct installation method.

Which of the following is the correct practice when installing two wires to a stud using flat washers and hex nuts?

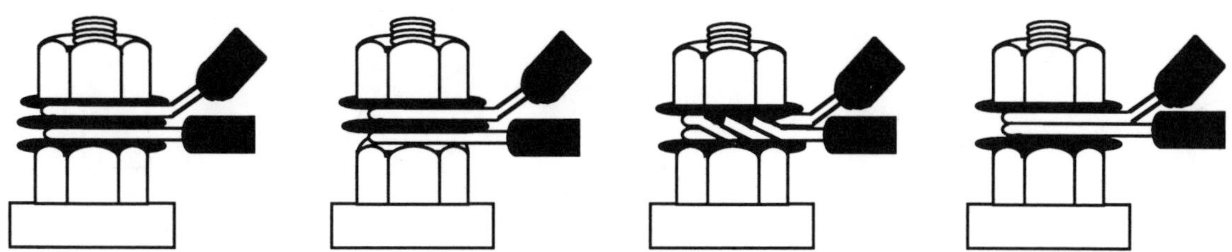

Which of the following is the correct practice when installing a wire around a binding post?

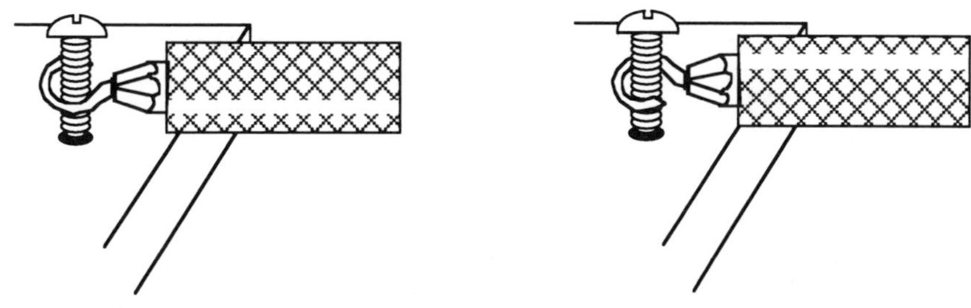

Fill in the blanks below naming the splices shown

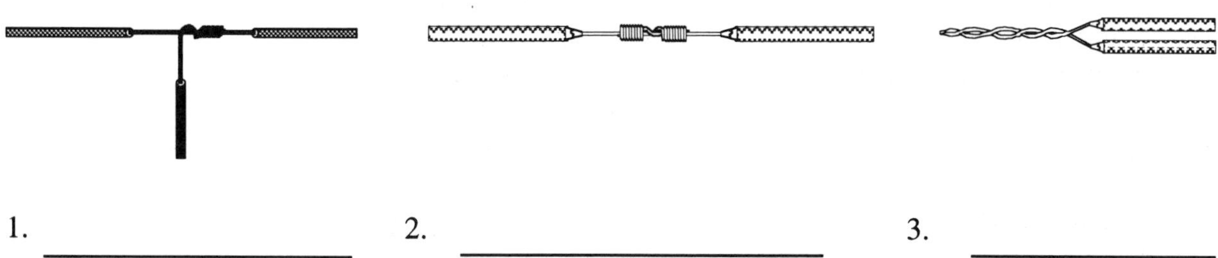

1. ____________________ 2. ____________________ 3. ____________________

JOURNEYMAN WIREMAN QUIZ #52
BLUEPRINT SYMBOLS

•Fill in the blank with the correct letter from choices below for the symbol

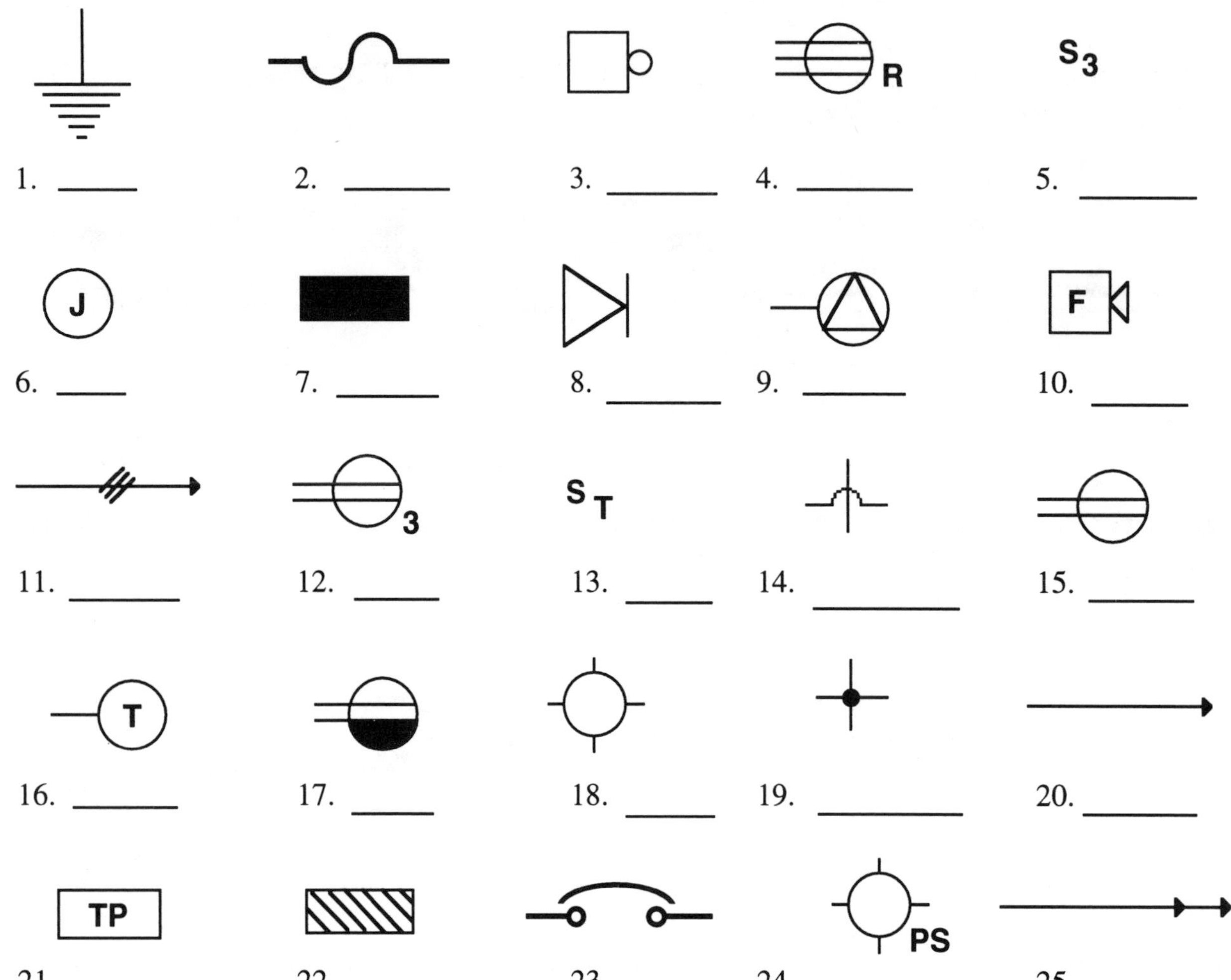

• *Choose a letter ( ) and fill in the blank above:*

(A) power panel
(B) fusible element
(C) two branch circuit home runs to panel
(D) time switch
(E) ceiling outlet
(F) duplex outlet, split circuit
(G) circuit breaker
(H) telephone
(I) thermostat
(J) fire alarm bell
(K) single special-purpose receptacle
(L) fire alarm horn
(M) wiring connected
(N) triplex receptacle outlet
(O) single branch circuit home run to panel
(P) wiring crossed not connected
(Q) lampholder with pull switch
(R) transformer pad
(S) junction box
(T) ground
(U) range receptacle
(W) switch 3-way
(X) duplex receptacle
(Y) branch circuit lighting panel
(Z) single branch circuit home run to panel (3-wire)

JOURNEYMAN QUIZ #53
TOOL IDENTIFICATION QUIZ

•*Fill in the blank with the correct letter from choices below for the tool shown:*

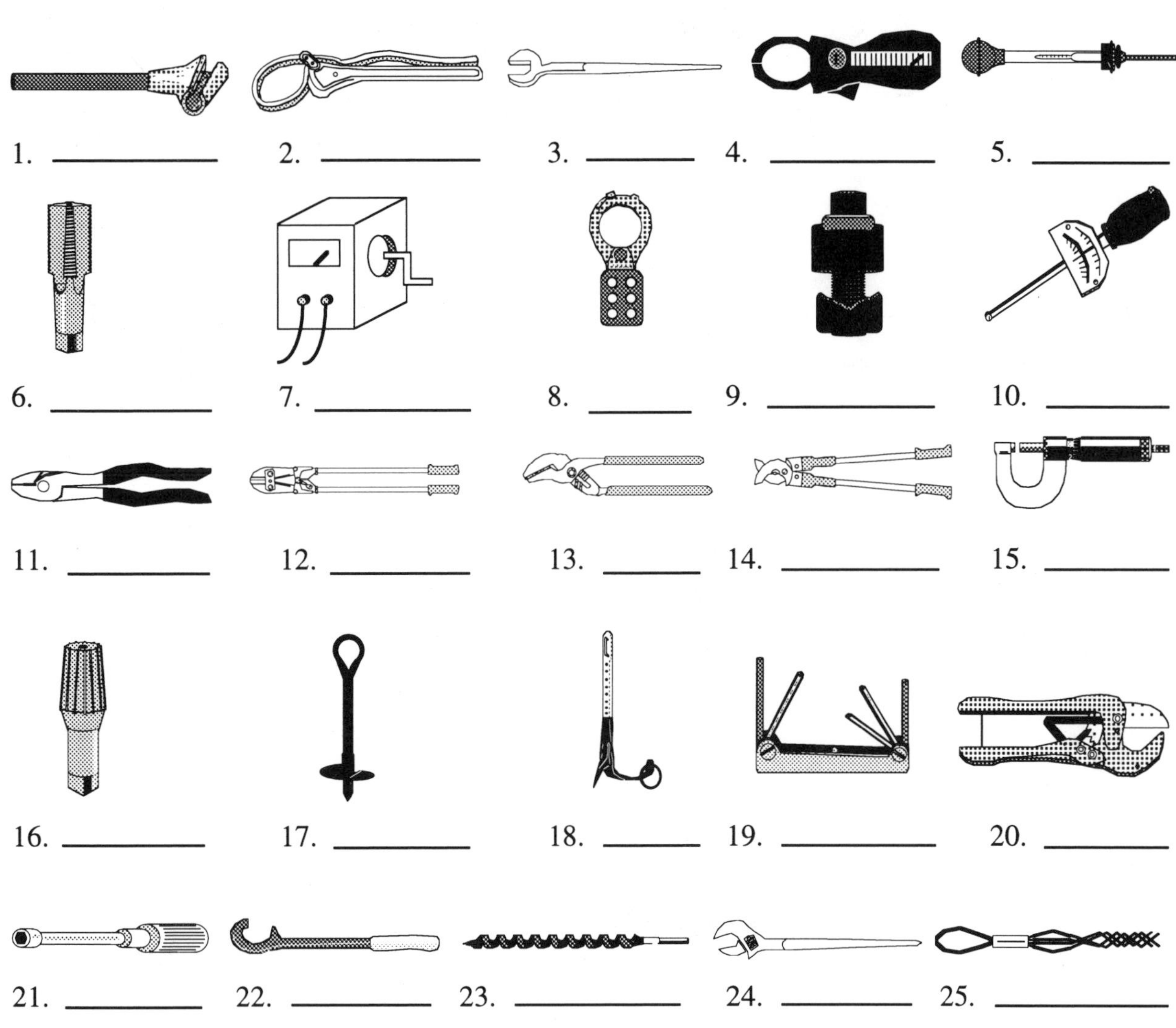

•*Choose a letter ( ) and fill in the blank above:*

(A) erection wrench
(B) hex key set
(C) strap wrench
(D) auger drill bit
(E) megger
(F) pipe thread tap
(G) wire mesh cable grip
(H) torque wrench
(I) bolt cutters
(J) construction wrench
(K) lock out
(L) pole climbers
(M) pipe reamer
(N) nut driver
(O) clamp on meter
(P) cable cutters
(Q) hydrometer
(R) hickey
(S) side cutting pliers
(T) guy anchor
(U) PVC cutters
(W) micrometer
(X) knock out
(Y) pump pliers
(Z) cable bender

JOURNEYMAN WIREMAN QUIZ #54
WIRING METHODS QUIZ

• *Fill in the blank with the correct letter for the wiring method or equipment:*

1. ______________

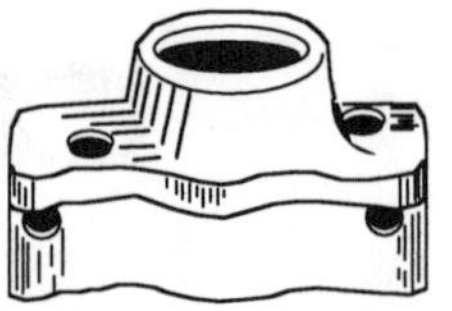

2. __________

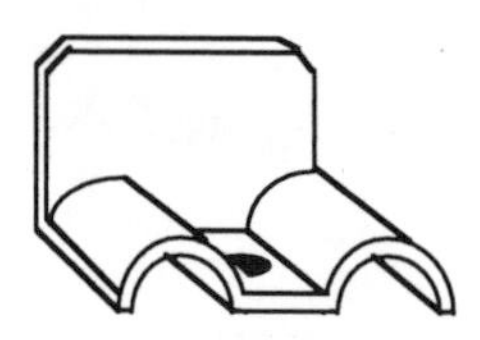

3. __________

4. ________

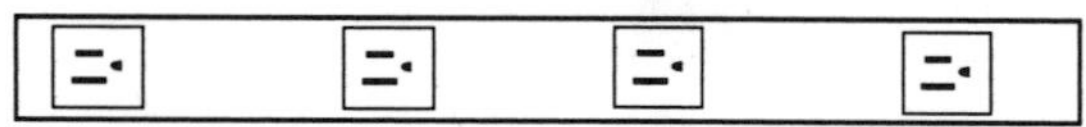

5. ______________________________

6. __________

7. ________

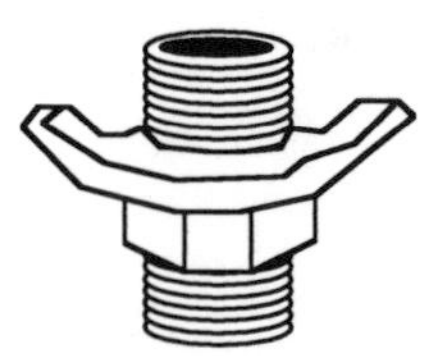

8. __________

9. ______

10. ______

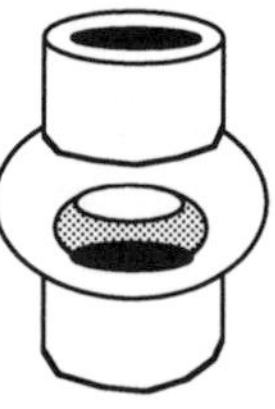

11. _______

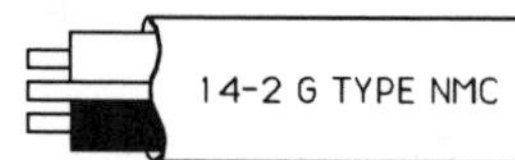

12. __________

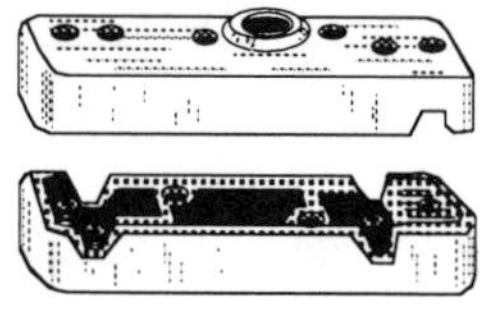

13. _________

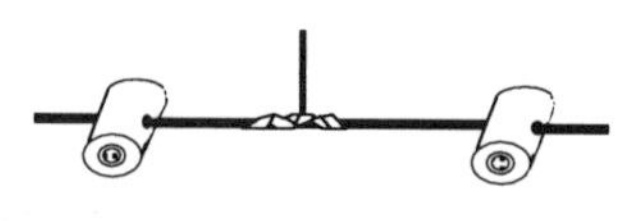

14. ______________

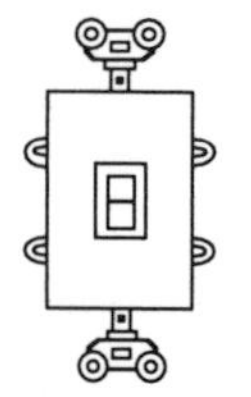

15. _______

• *Choose a letter ( ) and fill in the blank above:*

(A) hickey
(B) double-pole switch
(C) rosette
(D) open-wiring
(E) conduit tee
(F) fixture stud
(G) two-screw connector
(H) split-bolt connector
(I) Romex
(J) armored cable
(K) four-way switch
(L) wire nut
(M) cable clamp
(N) plugmold
(O) open-wiring receptacle

JOURNEYMAN WIREMAN QUIZ #55
BLUEPRINT SYMBOLS

•Fill in the blank with the correct letter from choices below for the symbol

S_K

1. ______

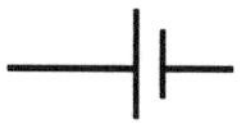

2. ______

WP

3. ______

F

4. ______

S

5. ______

S_P

6. ______

7. ______

D

8. ______

M

9. ______

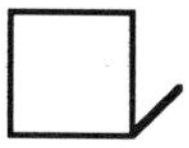

10. ______

11. ______

S_F

12. ______

13. ______

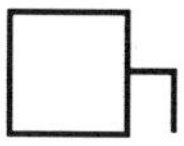

14. ______

H

15. ______

16. ______

R

17. ______

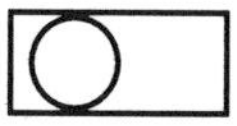

18. ______

B

19. ______

F

20. ______

21. ______

GR

22. ______

23. ______

24. ______

• Choose a letter ( ) and fill in the blank above:

(A) controller
(B) motor
(C) bare-lamp fluorescent strip
(D) motor starter
(E) electric door opener
(F) switch fused
(G) fan outlet
(H) push button
(I) fire alarm bell
(J) buzzer
(K) one cell
(L) safety switch
(M) fixture recessed
(N) weatherproof outlet
(O) ceiling pull switch
(P) blanked outlet
(Q) grounded duplex receptacle
(R) wiring or conduit turned down
(S) switch with pilot
(T) fluorescent fixture
(U) heating panel
(W) switch key operated
(X) humidistat
(Y) wiring or conduit turned up

JOURNEYMAN QUIZ #56
TOOL IDENTIFICATION QUIZ

•*Fill in the blank with the correct letter from choices below for the tool shown:*

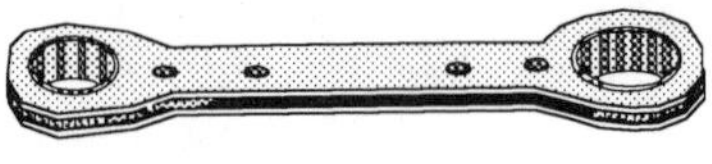

1. 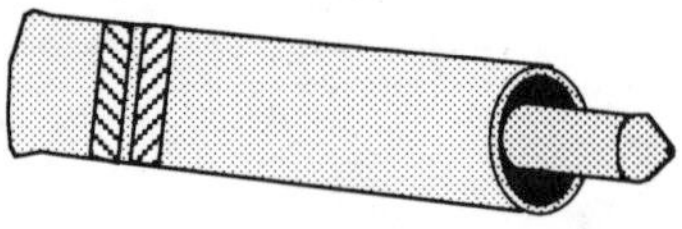 ___________ 2. 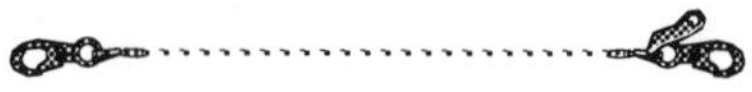 ___________ 3. ___________

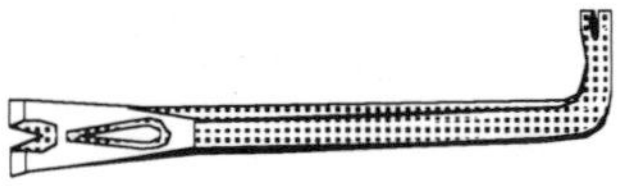

4.  ___________ 5.  ___________ 6. ___________

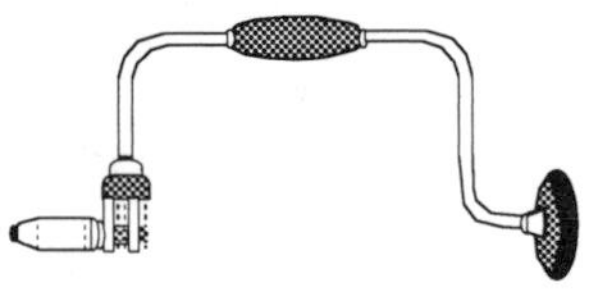

7.  ___________ 8. 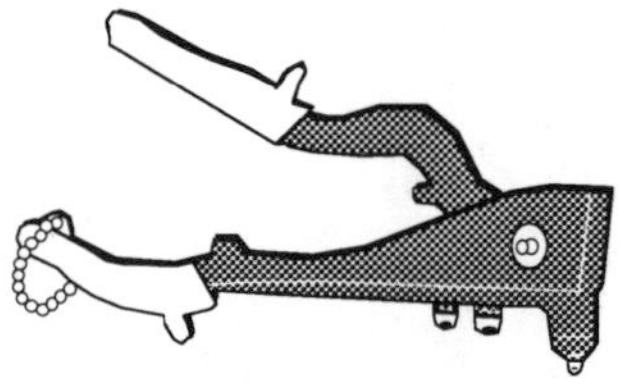 ___________ 9. ___________

10. 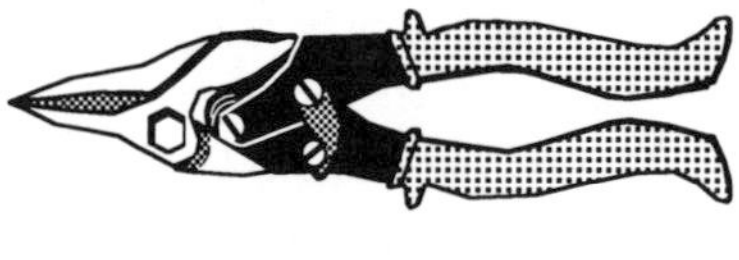 ___________ 11. 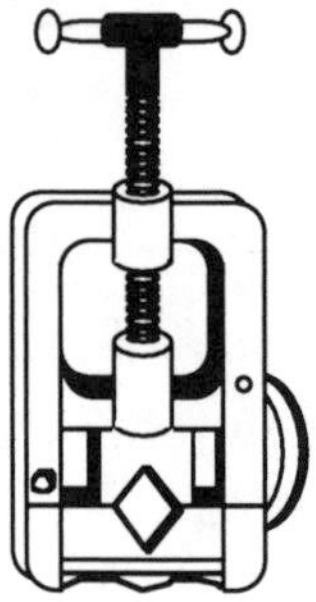 ___________ 12. ___________

• *Choose a letter ( ) and fill in the blank above:*

(A) box end wrench	**(G)** pipe vise	**(M)** bit brace
(B) fish-tape leader	**(H)** center punch	**(N)** auger bit
(C) ripping bar	**(I)** drift punch	**(O)** ratcheting box wrench
(D) hydrometer	**(J)** rotometer	**(P)** lead anchor set
(E) aviation snips	**(K)** pipe clamp	**(Q)** star drill
(F) lineman pliers	**(L)** rope lanyard	**(R)** riveting tool

JOURNEYMAN WIREMAN QUIZ #57
WIRING METHODS QUIZ

• *Fill in the blank with the correct letter for the wiring method or equipment:*

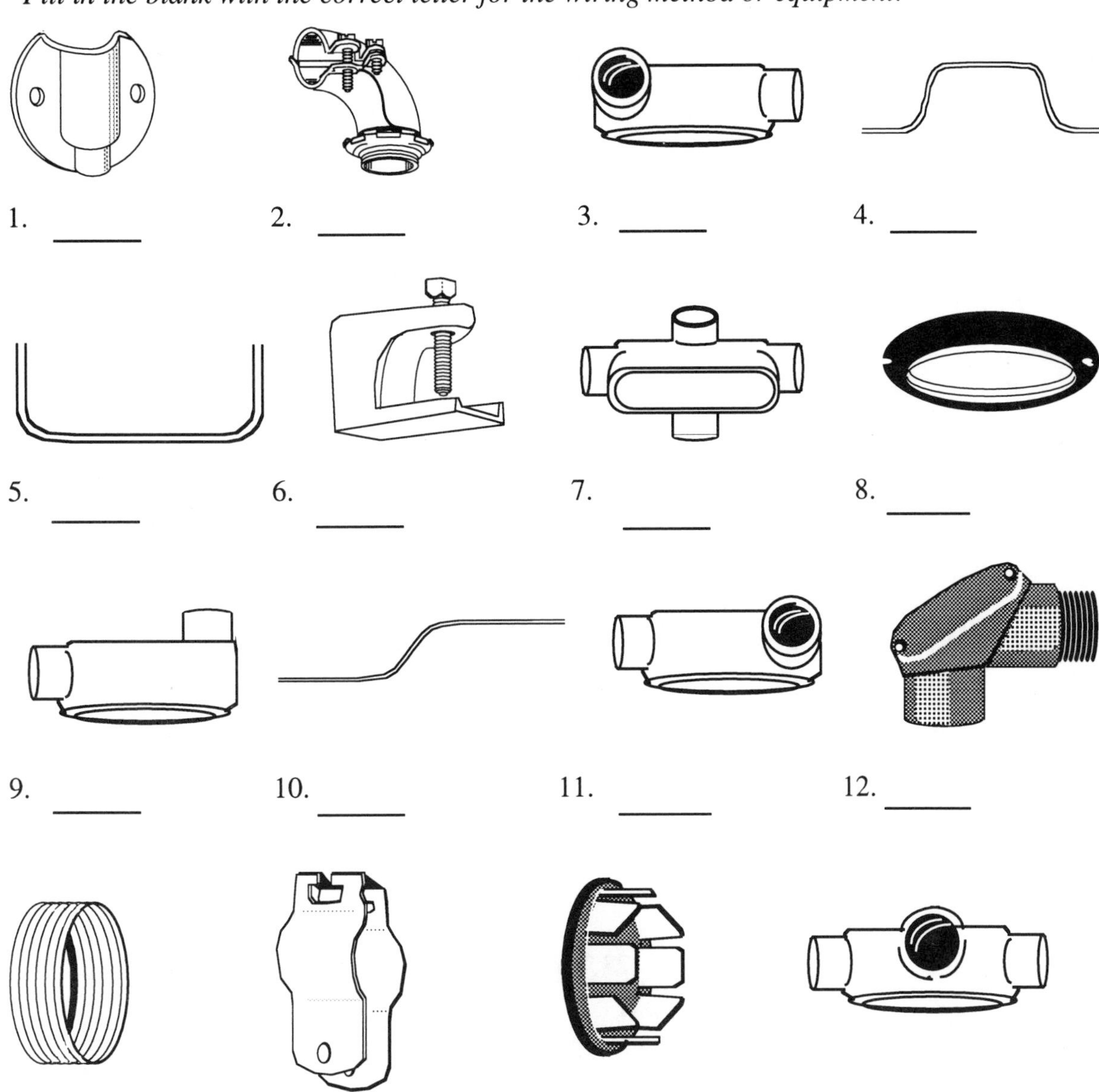

• *Choose a letter ( ) and fill in the blank above:*

(A) T conduit body
(B) plaster ring
(C) beam clamp
(D) back-to-back bend
(E) conduit hanger
(F) LB conduit body
(G) pulling 90° elbow
(H) knockout blank
(I) LL conduit body
(J) dog leg bend
(K) armored cable connector
(L) X conduit body
(M) cable sill plate
(N) reducing bushing
(O) saddle bend
(P) LR conduit body

JOURNEYMAN WIREMAN QUIZ #58
BLUEPRINT SYMBOLS

•*Fill in the blank with the correct letter from choices below for the symbol*

WH CH C

1.  _____ 2.  _____ 3.  _____ 4. 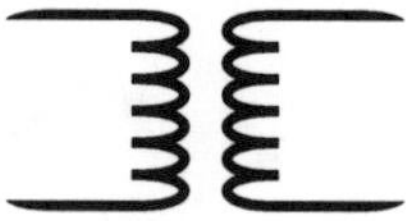 _____ 5. _____

S 4 TV

6. _____ 7. 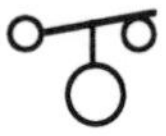 _____ 8.  _____ 9.  _____ 10. _____

11.  _____ 12.  _____ 13.  _____ 14.  _____ 15. _____

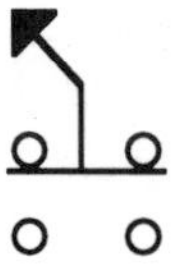

16.  _____ 17.  _____ 18. 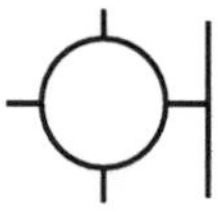 _____ 19.  _____ 20. _____

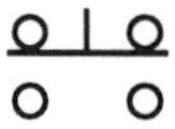

21. 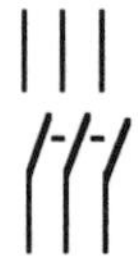 _____ 22. 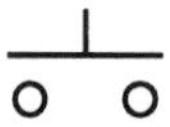 _____ 23.  _____ 24. 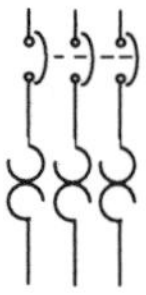 _____ 25. _____

• *Choose a letter ( ) and fill in the blank above:*

(A) temperature actuated switch N.C.
(B) normally open foot switch
(C) phase
(D) wall bracket
(E) chime
(F) television outlet
(G) disconnect
(H) switch four-way
(I) thermal O.L.
(J) watthour meter
(K) liquid level switch N.C.
(L) SPDT double break
(M) mushroom head push button
(N) transformer
(O) CB with thermal O.L.
(P) delta
(Q) clock
(R) stop button N.C.
(S) N.O. contact
(T) timed contact N.O.T.C.
(U) limit switch N.O.
(W) start button N.O.
(X) selector switch two-position
(Y) N.C. contact
(Z) floor outlet

JOURNEYMAN QUIZ #59
TOOL IDENTIFICATION QUIZ

•Fill in the blank with the correct letter from choices below for the tool shown:

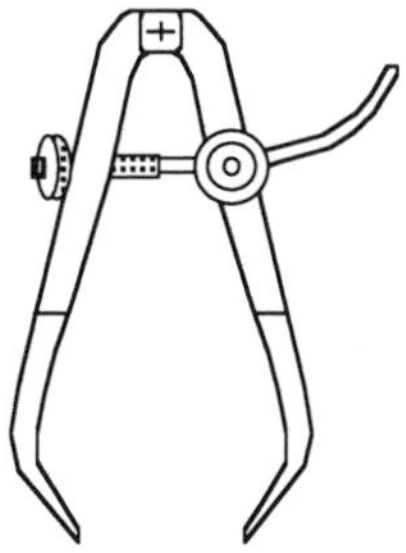

1.  ____________ 2. 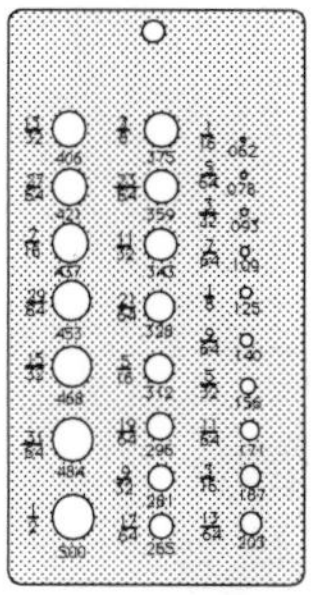 ____________ 3. ____________

4. 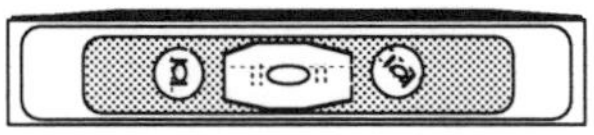 ____________ 5. 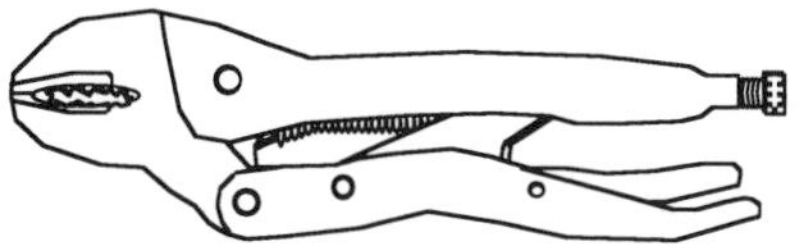 ____________ 6. ____________

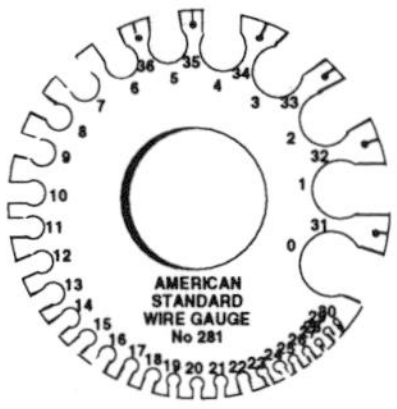

7. 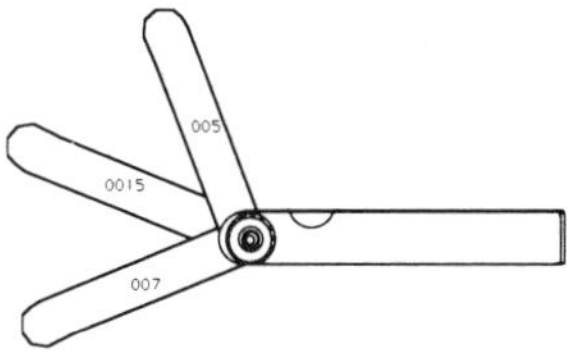 ____________ 8. 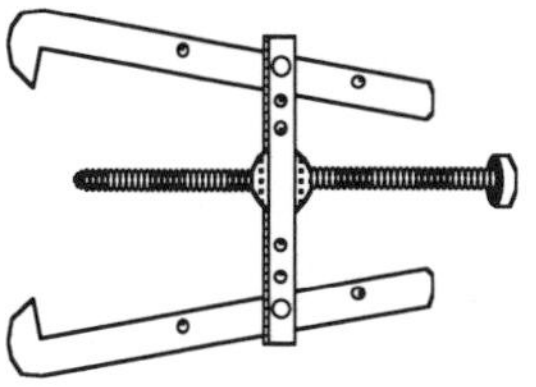 ____________ 9. ____________

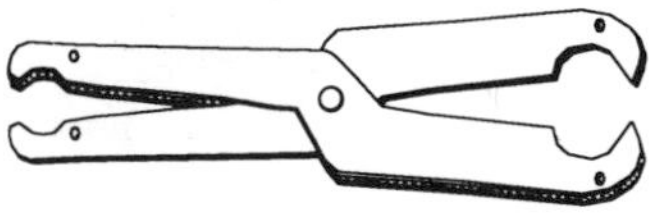

10.  ____________ 11. 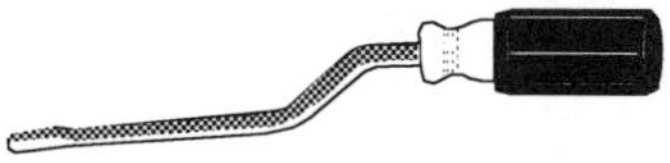 ____________ 12. ____________

• Choose a letter ( ) and fill in the blank above:

(A) scratch awl
(B) expansion bit
(C) pump pliers
(D) feeler gauge
(E) fuse puller
(F) pipe holder
(G) drill bit gauge
(H) rotary screwdriver
(I) hole saw
(J) wire gauge
(K) offset screwdriver
(L) scribe
(M) caliper
(N) gear puller
(O) depth checker
(P) torpedo level
(Q) plumb bob
(R) locking pliers

JOURNEYMAN WIREMAN QUIZ #60
WIRING METHODS QUIZ

• *Fill in the blank with the correct letter for the wiring method or equipment:*

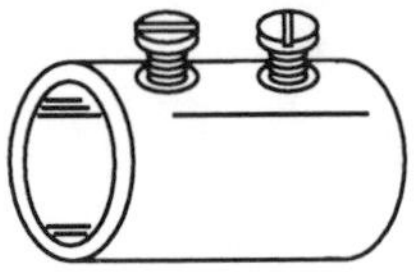

1. 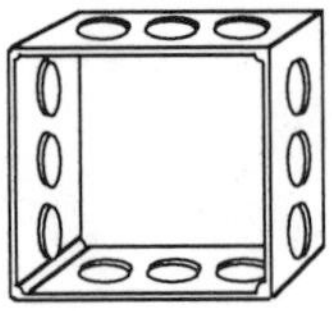 ________ 2. 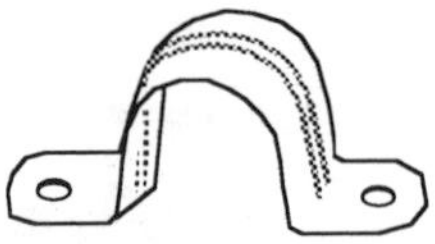 ________ 3. 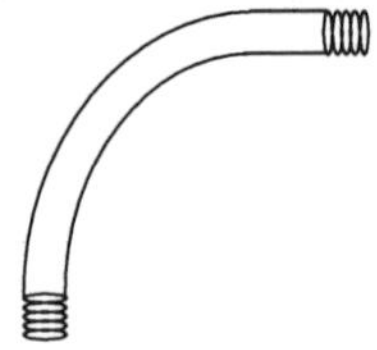 ________ 4. ________

5. 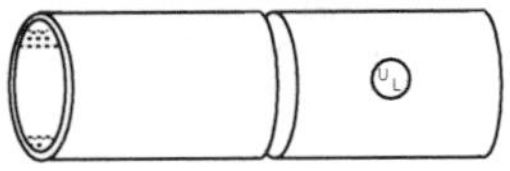 ________ 6.  ________ 7. 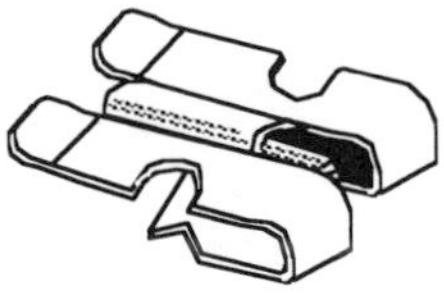 ________ 8. ________

9. 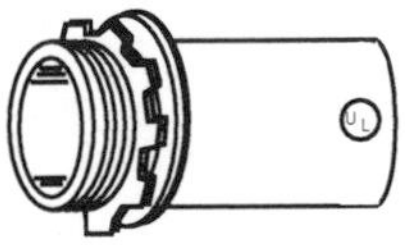 ________ 10.  ________ 11. 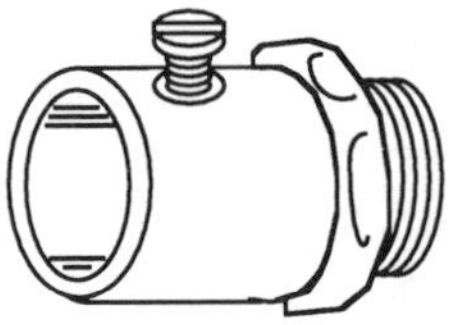 ________ 12. ________

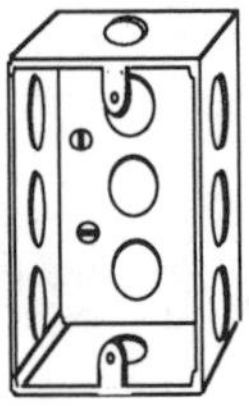

13. 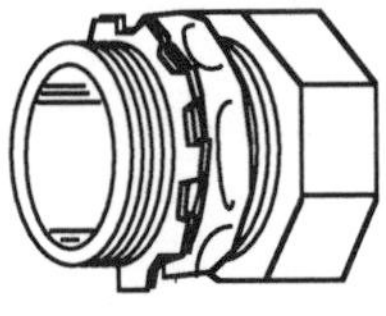 ________ 14.  ________ 15.  ________ 16. ________

• *Choose a letter ( ) and fill in the blank above:*

(A) ring tongue terminal
(B) octagon box
(C) set screw type connector
(D) 50 amp receptacle
(E) indenter type connector
(F) grounding clip
(G) oval service cable strap
(H) butt connector
(I) sweeping elbow
(J) 30 amp receptacle
(K) compression type connector
(L) indenter type coupling
(M) square box
(N) set screw type coupling
(O) handy box
(P) pipe strap

JOURNEYMAN WIREMAN QUIZ #61
PRACTICAL QUESTIONS

•*Circle the correct answer:*

1. If the Line 1 fuse is blown and Line 2 fuse is okay, the test light that will be lit is # ____.

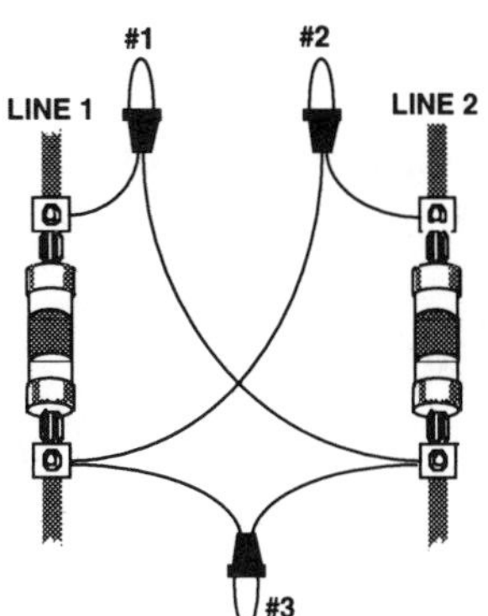

(a) #1 (b) #2 (c) #3 (d) #1 and #2

2. A recommended safe distance between the foot of an extension ladder and the wall that the ladder is placed against is _____.

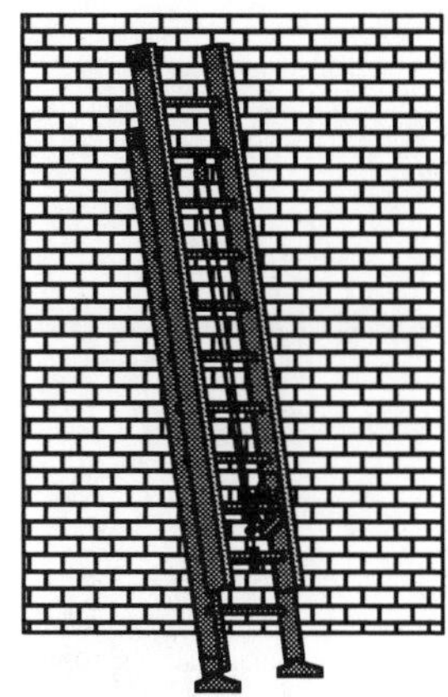

(a) 1/2 the length of the extended ladder
(b) no more than 3 feet
(c) 1/4 the length of the extended ladder
(d) 5 feet for ladders less than 16 feet

3. Which wires are connected together so the switch will control both lights?

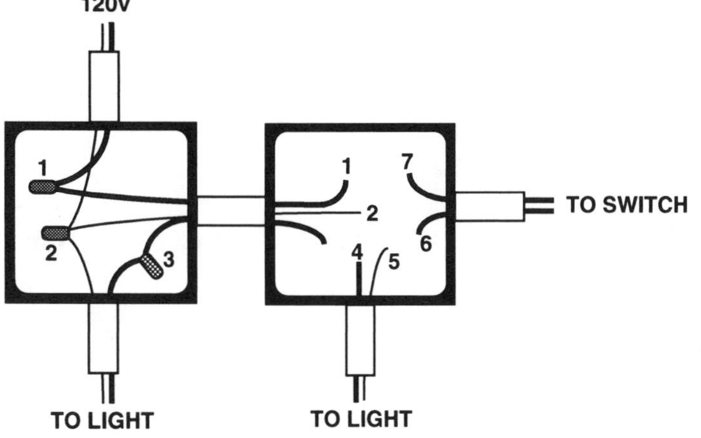

(a) 1 to 5, 2 to both 6 and 7, 3 to 4

(b) 1 to 7, 2 to both 5 and 6, 3 to 4

(c) 1 to 4, 2 to 6, 3 to both 5 and 7

(d) 1 to 6, 2 to 5, 3 to both 4 and 7

JOURNEYMAN WIREMAN QUIZ #62
PRACTICAL QUESTIONS

•Circle the correct answer:

1. A neon test light is connected across the terminals of a single-pole switch to a flourescent light on a 120 volt branch circuit. The test light will be lit when ____.

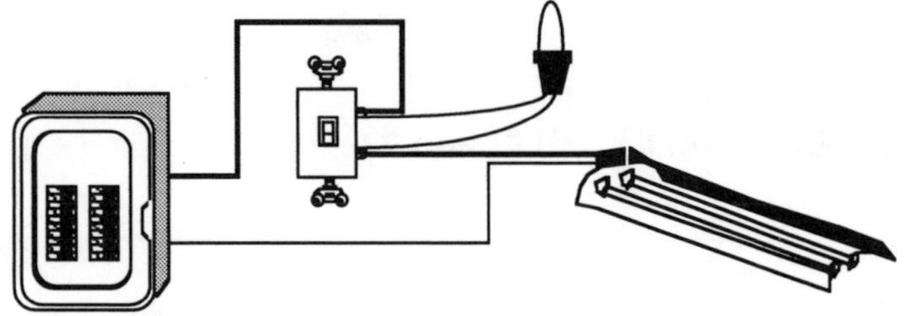

(a) the switch is turned on
(b) the switch is turned off
(c) both (a) and (b)
(d) neither (a) nor (b)

2.

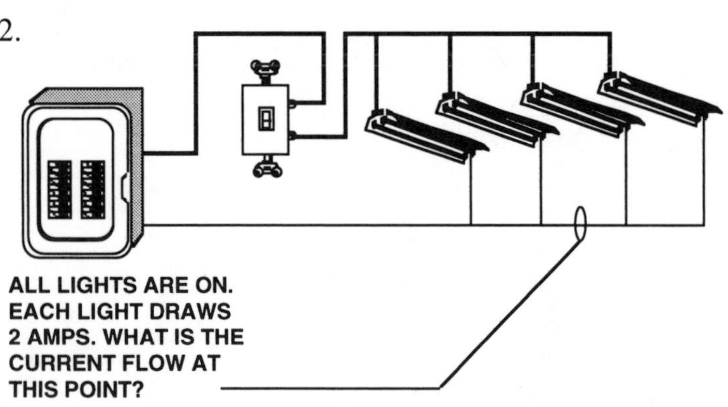

(a) 2 amps
(b) 4 amps
(c) 6 amps
(d) 8 amps

3. The meter will read ____ volts.

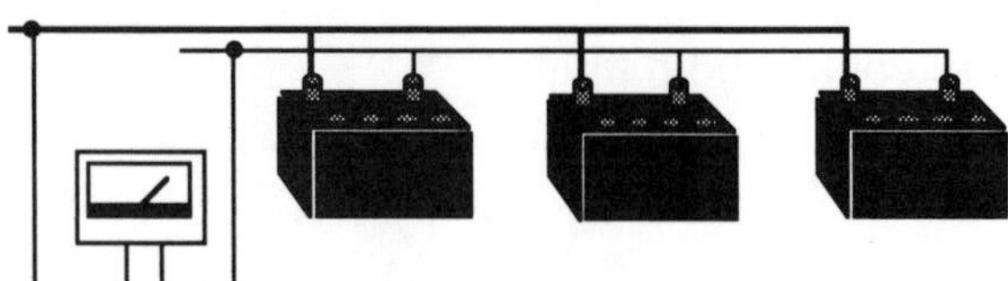

(a) 6 volts (b) 9 volts (c) 12 volts (d) 18 volts

EACH BATTERY IS 6 VOLTS

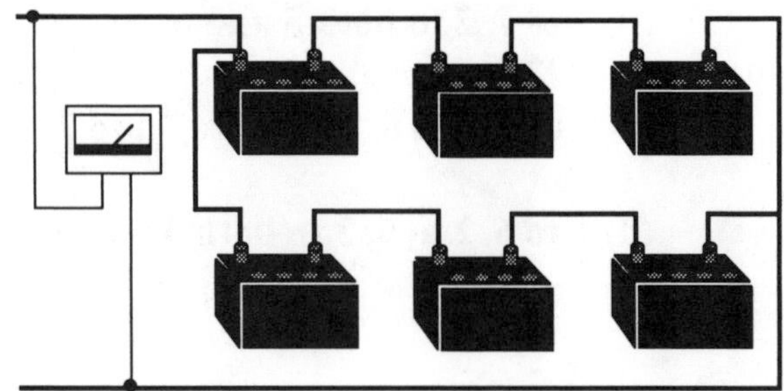

(a) 12 volts (b) 18 volts (c) 24 volts (d) 36 volts

FINAL
3 HOUR
EXAM

The following is a 3 hour exam that some areas use for the Journeyman license exam.

The exam is in 3 parts:

PARTS BOARD IDENTIFICATION
THEORY - OHMS LAW
CODE BOOK QUESTIONS

All 3 parts are closed book.

The Journeyman applicant will be given a parts board containing 36 items, you must be able to identify 20 of the items.

The second part of the exam you will have 20 questions from theory-Ohms law to answer.

The third part of the exam you will have 20 closed book Code questions to answer.

70% is passing. To grade your score count the number of correct answers and divide by the number of questions.

You have a total of 3 hours. You can spend as much or as little time on each part as you choose.

The following pages contain two complete exams. Work Exam #1 and grade yourself from the answer section before proceeding to Exam #2.

Remember, each Exam is 3 hours total, closed book.

EXAM #1

CLOSED BOOK

3 HOURS

PARTS BOARD EXAM #1

• *Fill in the blank with the correct letter shown below to identify the part.*

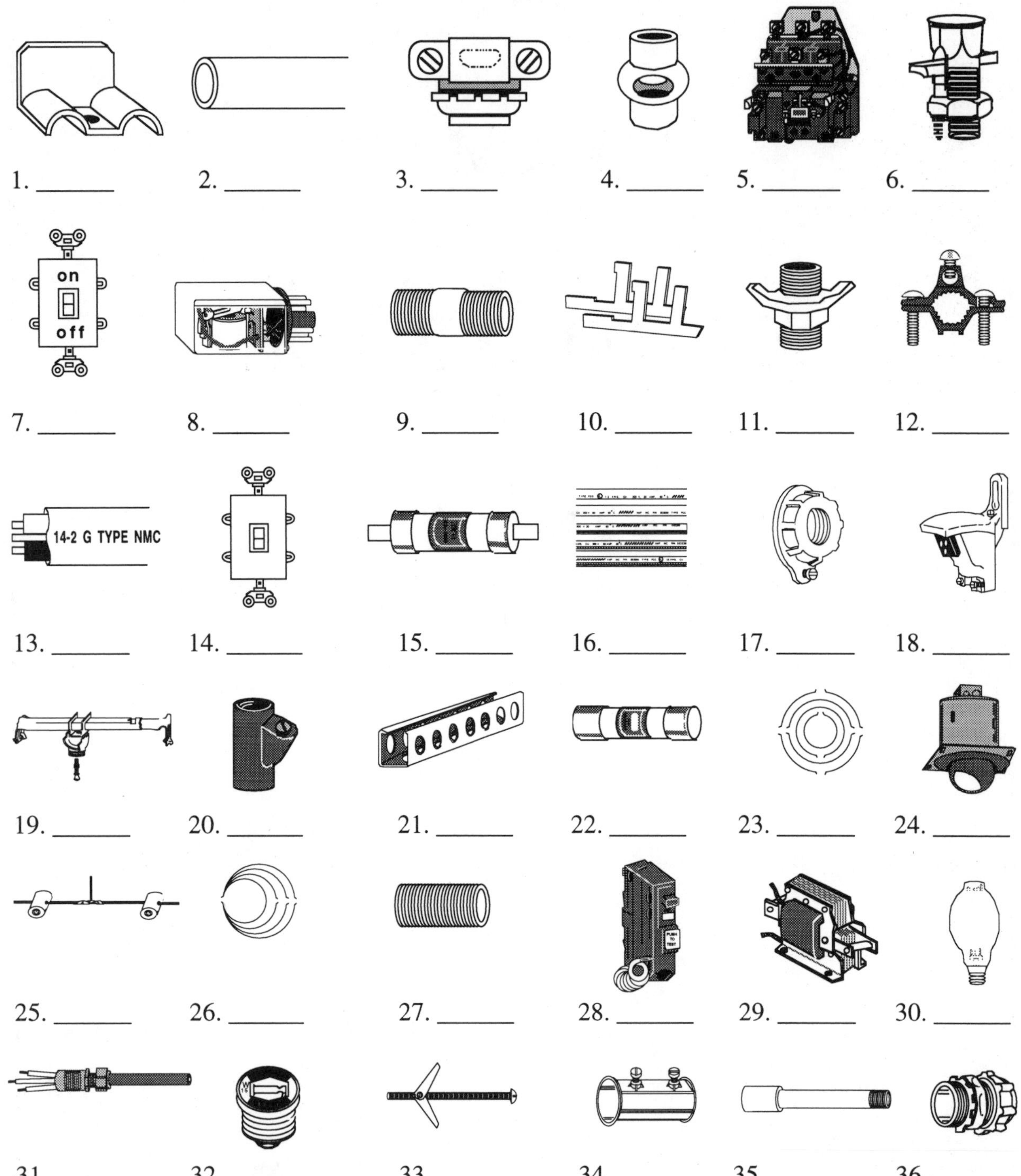

A. plug fuse B. EMT coupling C. grounding bushing D. cable clamp E. hickey F. two-screw connector G. fixture stud H. rigid conduit I. double-pole switch J. kindorf K. open wiring L. ground clamp M. Romex N. toggle bolt O. four-way switch P. knife fuse Q. FCC R. MI cable S. service head T. madison strap U. bar hanger V. close nipple W. eccentric KO X. concentric KO Y. bull's light Z. mercury lamp AA. seal off BB. GFCI receptacle CC. cartridge fuse DD. 3-way switch EE. BX cable FF. short nipple GG. EMT HH. relay II. split-bolt connector JJ. box connector KK. solenoid LL. GFCI breaker MM. pressure switch NN. motor starter

OHMS LAW - THEORY EXAM #1

1. A corroded electrical connection ____.

(a) decreases the voltage drop
(b) decreases the resistance of the connection
(c) increases the resistance at the connection
(d) increases the ampacity at the connection

2. A3Ω, 6Ω, 9Ω and a 12Ω resistor are connected in series. The resistor that will consume the most power is the ____ ohm.

(a) 3Ω (b) 6Ω (c) 9Ω (d) 12Ω

3. Resistance opposes the flow of current in a circuit and is measured in ____.

(a) farads (b) joules (c) ohms (d) henrys

4. A magnetic field is created around a conductor ____.

(a) whenever current flows in the wire, provided the wire is made of magnetic material
(b) only when the wire carries a large current
(c) whenever current flows in the conductor
(d) only if the conductor is formed into a loop

5. Which of the following is **not** true?

(a) Equal currents flow in the branches of parallel circuits.
(b) The total resistance of a parallel circuit is less than the smallest resistor in the circuit.
(c) The total current in a parallel circuit is the sum of the branch currents.
(d) In a parallel circuit, there is more than one path for the current flow.

6. In solving series-parallel circuits, generally you would ____.

(a) treat it as a series circuit
(b) reduce it to its simplest form
(c) assume that all loads are equal
(d) treat it as a parallel circuit

7. If a 240 volt heater is used on 120 volt, the amount of heat produced will be ____.

(a) twice as great (b) four times as great (c) 1/4 as much (d) 1/2 as much

8. The voltage of a circuit is best defined as ____.

(a) the potential between two conductors
(b) the greatest difference of potential between two conductors
(c) the effective difference of potential between two conductors
(d) the average RMS difference of potential between any two conductors

9. The advantage of AC over DC includes which of the following?

(a) better speed control
(b) lower resistance at higher current
(c) ease of voltage variation
(d) impedance is greater

10. When resistors are connected in series, the total resistance is ____.

(a) the sum of the individual resistance values
(b) the equivalent of the smallest resistance value
(c) the equivalent of the largest resistance value
(d) less than the value of the smallest resistance

11. If a 120 volt incandescent light bulb is operating at a voltage of 125 volts, the result will be ____.

(a) it may be enough to blow a fuse
(b) the bulb won't be as bright
(c) shorter life of the bulb
(d) the wattage will be less than rated

12. Four heaters, each having a resistance of 30 ohms, are connected in series across a 600-volt train circuit. The current is ____ amperes.

(a) 5 (b) 17 (c) 20 (d) 80

13. In a series circuit ____ is common.

(a) resistance (b) current (c) voltage (d) wattage

14. A wattmeter indicates ____.

I. real power   II. apparent power if PF is not in unity   III. power factor

(a) I only   (b) II only   (c) III only   (d) I, II and III

15. Two 500 watt lamps connected in series across a 110 volt line draws 2 amperes. The total power consumed is ____ watts.

(a) 50   (b) 150   (c) 220   (d) 1000

16. The resistance of a copper wire to the flow of electricity _____.

(a) decreases as the length of the wire increases
(b) decreases as the diameter of the wire decreases
(c) increases as the diameter of the wire increases
(d) increases as the length of the wire increases

17. A 15 ohm resistance carrying 20 amperes of current uses ____ watts of power.

(a) 300   (b) 3000   (c) 6000   (d) 9000

18. A common fuse and circuit breaker works on the principal that ____.

(a) voltage develops heat   (b) voltage breaks down insulation
(c) current develops heat   (d) current expands a wire

19. What is the total wattage of this circuit?

(a) 3.5   (b) 420   (c) 16,800   (d) 140

20. If the circuit voltage is increased, all else remains the same, only the ____ will change.

(a) resistance   (b) current   (c) ampacity   (d) conductivity

CODE BOOK QUESTIONS EXAM #1

1. Total load on any overcurrent device located in a panelboard shall not exceed ____ percent of its rating where, in normal operation, the load will continue for 3 hours or more.

(a) 80 (b) 100 (c) 60 (d) 50

2. When rigid metal conduits are buried the minimum cover required by the Code is ____.

(a) 6" (b) 12" (c) 18" (d) 24"

3. Identified, as used in the Code in reference to a conductor or its terminals, means that such a conductor or terminal is to be recognized as ____.

(a) grounded (b) bonded (c) colored (d) marked

4. In a residence, no point along the floor line in any wall space may be more than ____ feet from an outlet.

(a) 6 (b) 6 1/2 (c) 12 (d) 10

5. What is the minimum size fixture wire?

(a) #16 (b) #18 (c) #20 (d) #22

6. Ground rod electrodes shall be installed such that ____ of length is in contact with the soil.

(a) 6' (b) 7' (c) 7' 6" (d) 8'

7. The Code requires at least ____ inches of free conductor shall be left at each outlet and switch point.

(a) 6 (b) 8 (c) 10 (d) 12

8. Connection by means of wire binding screws or studs and nuts having upturned lugs or equivalent shall be permitted for ____ or smaller conductors.

(a) #10 (b) #8 (c) #6 (d) #4

9. Compliance with the provisions of the Code will result in ____.

(a) good electrical service (b) an efficient system (c) freedom from hazard (d) all of these

10. All of the following conductors can be connected in parallel **except** ____.

(a) #250 kcmil (b) #2/0 (c) #1 (d) #1/0

11. Which of the following electrodes must be supplemented by an additional electrode?

(a) metal underground water pipe **(b) metal frame of a building**
(c) ground ring **(d) concrete encased**

12. Fixed appliances rated at not over 300va or ____ hp the branch circuit overcurrent device shall be permitted to serve as the disconnecting means.

(a) 1/8 (b) 1/4 (c) 1/3 (d) 1/2

13. Plaster, drywall or plasterboard surfaces that are broken or incomplete shall be repaired so there will be no gaps or open spaces greater than ____ inch at the edge of the fitting box.

(a) 1/16 (b) 1/8 (c) 3/16 (d) 1/4

14. What is the minimum height of a service drop attachment to a building?

(a) 8 feet (b) 10 feet (c) 12 feet (d) 15 feet

15. Receptacles located within ____ feet of the inside wall of a pool shall be protected by a GFCI.

(a) 8 (b) 10 (c) 15 (d) 20

16. What is the minimum size conductor permitted for general wiring under 600 volts?

(a) #12 copper (b) #14 aluminum (c) #14 copper (d) #12 aluminum

17. The maximum number of 90° quarter bends in one run of EMT is ____.

(a) two (b) four (c) five (d) three

18. The volume per #14 conductor required in box sizing is ____ cubic inch.

(a) 2.25 (b) 2 (c) 2.5 (d) 3

19. Temporary wiring shall be removed ____ upon completion of construction or purpose for which the wiring was installed.

(a) 30 days (b) immediately (c) A.S.A.P. (d) 60 days

20. In completed installations each outlet box shall have a ____.

(a) receptacle (b) switch (c) cover (d) fixture

EXAM #2

CLOSED BOOK

3 HOURS

PARTS BOARD EXAM #2

• *Fill in the blank with the correct letter shown below to identify the part.*

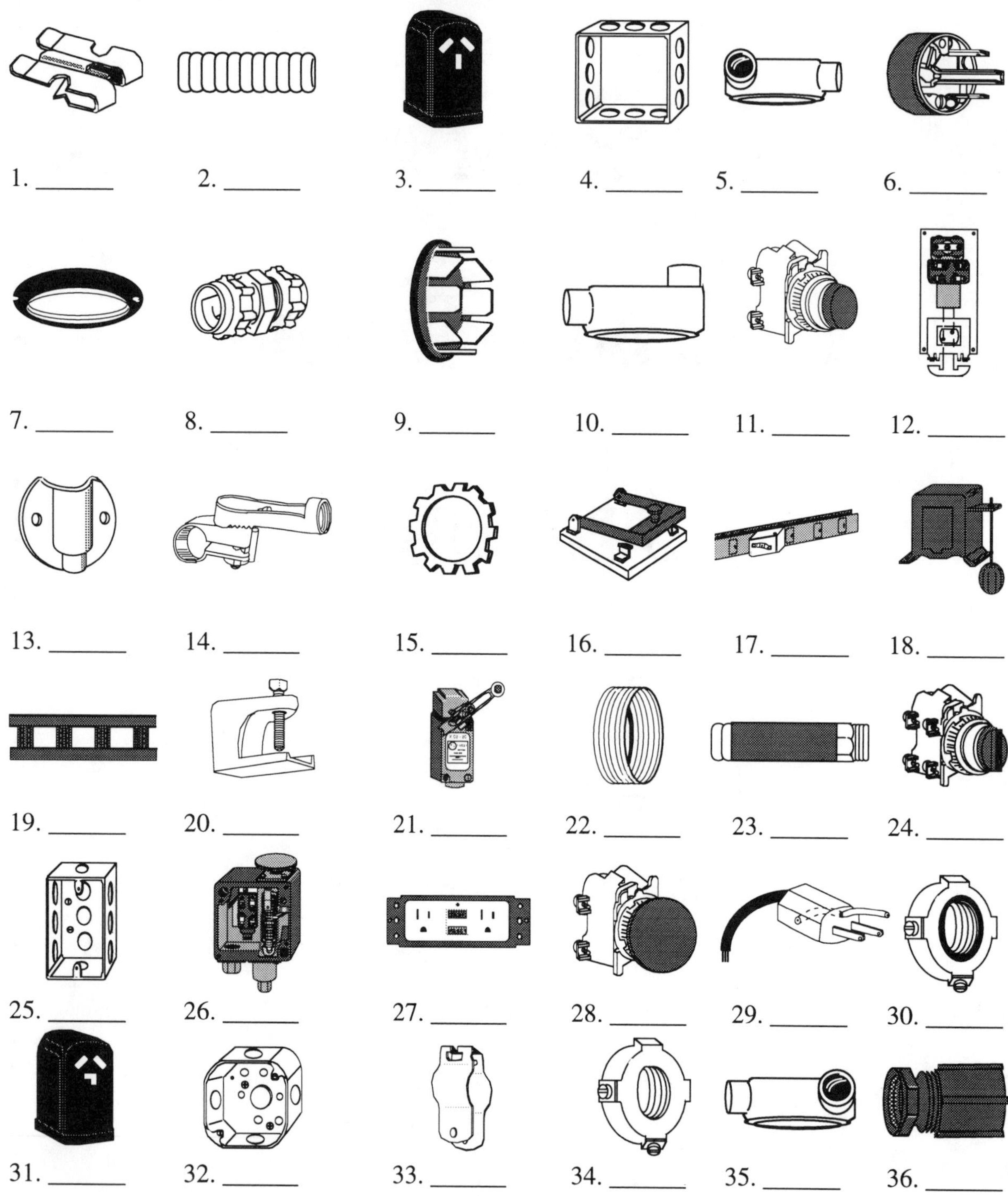

A. erickson B. conduit hanger C. limit switch D. liquid tight flex E. 30 amp receptacle F. knife switch G. knockout blank H. square box I. EMT coupling J. clip clamp K. BX L. selector switch M. attachment plug N. self-restoring plug O. LB P. stop button Q. locknut R. busway S. pressure switch T. 50 amp receptacle U. beam clamp V. handy box W. LL X. plaster ring Y. insulated throat bushing Z. reducing bushing AA. float switch BB. GFCI receptacle CC. octagon box DD. LR EE. start button FF. noninsulated throat bushing GG. timer HH. cable tray II. cable sill plate JJ. pipe ground connecter KK. solenoid LL. GFCI breaker MM. pressure switch

OHMS LAW - THEORY EXAM #2

1. A 3Ω, a 6Ω, a 9Ω and a 12Ω resistor are connected in parallel. Which resistor will consume the most power?

(a) 3Ω (b) 6Ω (c) 9Ω (d) 12Ω

2. One of the essential functions of any switch is to maintain a ____.

(a) good high-resistance contact in the closed position
(b) good low-resistance contact in the closed position
(c) good low-resistance contact in the open position
(d) good high-resistance contact in the open position

3. With switches 1 and 2 closed the combined resistance of the circuit is ____ ohms.

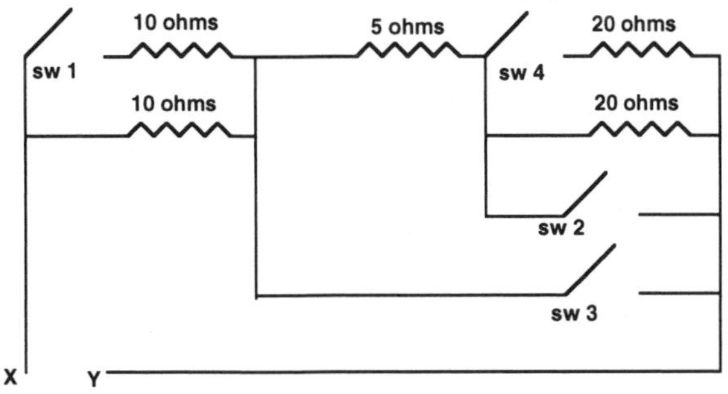

(a) 30 (b) 25 (c) 10 (d) 3

4. An electron is ____.

(a) a neutron **(b) an orbiting particle**
(c) a proton **(d) the smallest part of an atom with a negative charge**

5. A battery operates on the principle of ____.

(a) photo emission **(b) triboelectric effect**
(c) electro chemistry **(d) voltaic conductivity**

6. A shunt is used to measure _____.

(a) resistance (b) capacitance (c) current (d) wattage

7. When an electric current is forced through a wire that has considerable resistance, the ____.

I. ampacity will decrease II. voltage will drop III. wire will heat up

(a) III only (b) I and II only (c) II and III only (d) I and III only

8. Alternating currents may be increased or decreased by means of a ____.

(a) motor (b) transformer (c) dynamo (d) megger

9. A(an) ____ changes AC to DC.

(a) battery (b) capacitor (c) alternator (d) rectifier

10. ____ is the ability of a material to permit the flow of electrons.

(a) Voltage (b) Current (c) Resistance (d) Conductance

11. Using 1.5 volt dry cells, the voltage between A and B would be ____ volts.

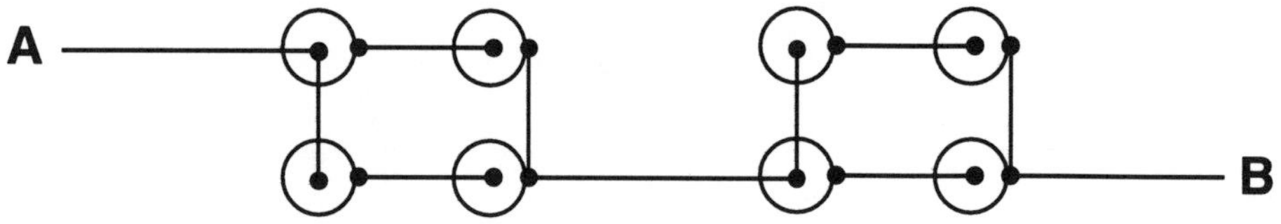

A

B

(a) 1.5 (b) 4 (c) 6 (d) 12

12. A toaster will produce less heat on low voltage because ____.

(a) its total watt output decreases **(b) the current will decrease**
(c) the resistance has not changed **(d) all of these**

13. If two equal resistance conductors are connected in parallel, the resistance of the two conductors is equal to ____.

(a) the resistance of one conductor
(b) twice the resistance of one conductor
(c) one-half the resistance of one conductor
(d) the resistance of both conductors

14. Which of the following is **not** the force which moves electrons?

(a) EMF (b) voltage (c) potential (d) current

15. If a low resistance is connected in parallel with a higher resistance, the combined resistance is ____.

(a) higher or lower than the low resistance depending on the size of the higher resistance
(b) always less than the low resistance
(c) always more than the higher resistance
(d) the total would be the low and high added together

16. The voltage drop in a line can be decreased by ____.

I. increasing the wire size
II. increasing the current
III. decreasing the load

(a) I only (b) I and II only (c) I, II and III (d) I and III only

17. When voltage and current appear at their zero and peak values at the same time, they are in ____.

(a) motion (b) group (c) phase (d) balanced

18. Which of the following statements is **incorrect**?

(a) current flowing through a conductor causes heat
(b) the conduit of an electrical system should be grounded
(c) volt meters are connected in parallel in a circuit
(d) rectifiers change DC to AC

19. The total resistance of four 10 ohm resistors in parallel is ____.

(a) 10 ohms (b) 2.5 ohms (c) 5 ohms (d) 4 ohms

20. The property of a circuit tending to prevent the flow of current and at the same time causing energy to be converted into heat is referred to as ____.

(a) the inductance (b) the resistance (c) the capacitance (d) the reluctance

CODE BOOK QUESTIONS EXAM #2

1. In a residence, hallways of ____ feet or more in length shall have at least one receptacle outlet.

(a) 4 (b) 6 (c) 8 (d) 10

2. In general, switches shall be so wired that all switching is done in the ____ conductor.

(a) grounded (b) ungrounded (c) grounding (d) ground

3. Insulated conductors smaller than ____, intended for use as grounded conductors of circuits, shall have an outer identification of white or gray color.

(a) #4 (b) #2 (c) #1/0 (d) 250 kcmil

4. The ampacity for conductors is derated when the ambient temperature exceeds ____.

(a) 30°F (b) 72°F (c) 86°F (d) 104°F

5. The primary purpose for grounding a raceway is to prevent the raceway from becoming ____.

(a) accidentally energized at a higher potential than ground
(b) a source of induction
(c) magnetized
(d) a path for eddy currents

6. Concrete, brick or tile walls are considered as being ____.

(a) isolated (b) insulators (c) grounded (d) dry locations

7. The location of a wall receptacle outlet in the bathroom of a dwelling shall be installed ____.

(a) the Code does not specify the location
(b) adjacent to the toilet
(c) adjacent to the basin
(d) across from the shower

8. A fixture that weighs more than ____ pounds shall be supported independently of the outlet box.

(a) 25 (b) 30 (c) 35 (d) 50

9. Is it permissible to install DC and AC conductors in the same outlet box?

(a) yes, if insulated for the maximum voltage of any conductor **(b) no, never**
(c) yes, if the ampacity is the same for both conductors **(d) yes, in dry places**

10. Listed ceiling (paddle) fans that do not exceed ____ pounds in weight, with or without accessories, shall be permitted to be supported by outlet boxes identified for such use.

(a) 35 (b) 45 (c) 50 (d) 60

11. Service drop conductors not in excess of 600 volts shall have a minimum clearance of ____ feet over residential property and driveways, and those commercial areas not subject to truck traffic.

(a) 10 (b) 12 (c) 15 (d) 18

12. The connection of a ground clamp to a grounding electrode shall be ____.

(a) accessible (b) visible (c) readily accessible (d) in sight

13. Mandatory rules of the Code are identified by the use of the word ____.

(a) should (b) shall (c) must (d) could

14. _____ lighting is a string of outdoor lights suspended between two points more than 15' apart.

(a) Pole (b) Festoon (c) Equipment (d) Outline

15. The Code considers low voltage to be ____.

(a) 480 volts or less (b) 600 volts or less (c) 24 volts (d) 12 volts

16. All wiring must be installed so that when completed ____.

(a) it meets the current-carrying requirements of the load
(b) it is free of shorts and unintentional grounds
(c) it is acceptable to Code compliance authorities
(d) it will withstand a hy-pot test

17. When mounting electrical equipment, wooden plugs driven into holes in ____ shall **not** be used.

I. masonry II. concrete III. plaster

(a) I only (b) II only (c) III only (d) I, II or III

18. The Code is designed for safety regardless of ____.

I. cost II. time III. maintenance IV. efficiency V. future expansion

(a) I and II (b) III and IV (c) I through IV (d) I through V

19. The maximum number of overcurrent devices that may be installed in a lighting panel is ____.

(a) 24 (b) 36 (c) 42 (d) 48

20. The path to ground from circuits, equipment, and conductor enclosures shall ____.

I. have sufficiently low impedance to limit the voltage to ground and to facilitate the operation of the circuit protective devices in the circuit
II. have capacity to conduct safely any fault current likely to be imposed on it
III. be permanent and continuous

(a) I only (b) II only (c) III only (d) I, II and III

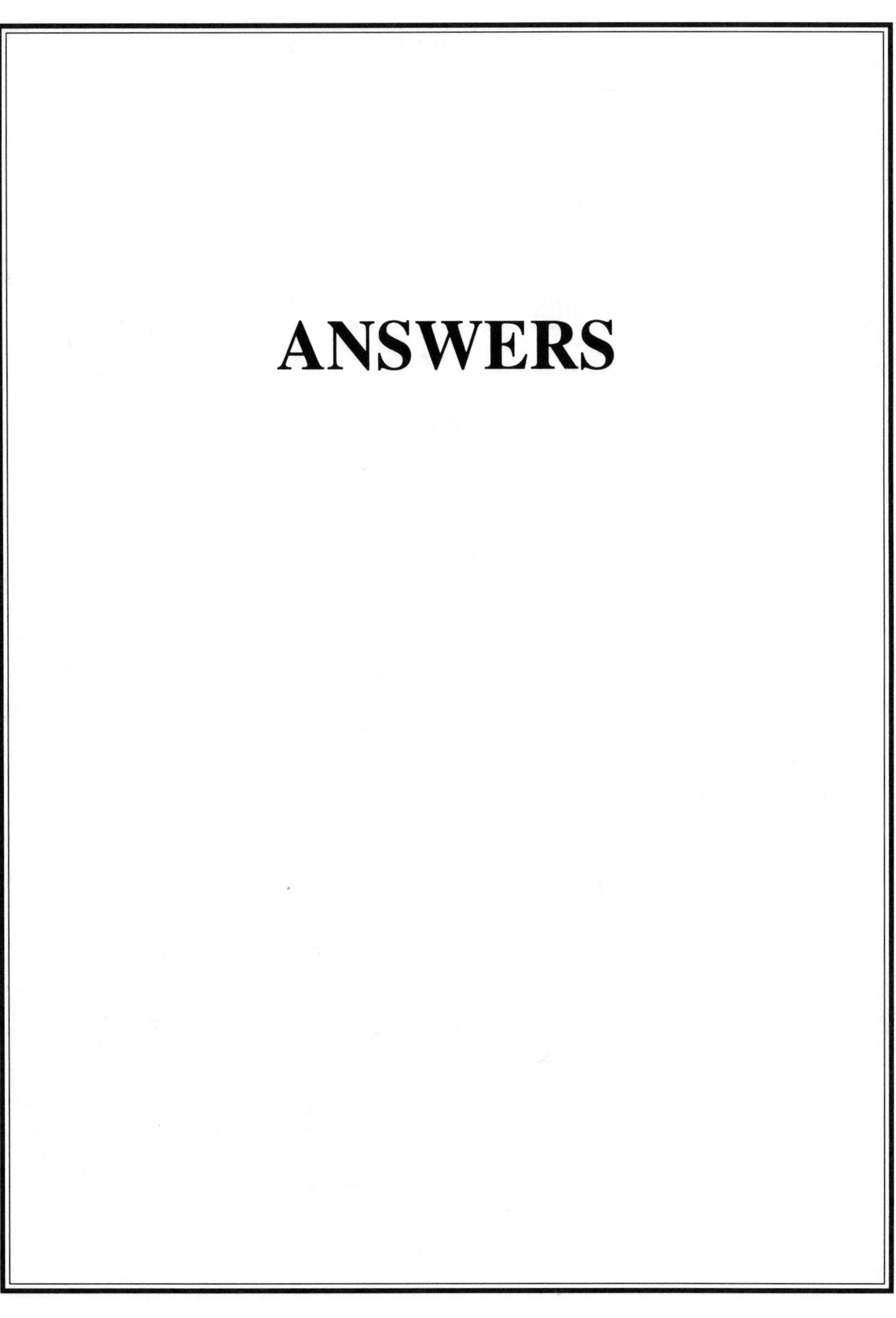

ANSWERS

JOURNEYMAN WIREMAN QUIZ #1
BLUEPRINT SYMBOLS

•Fill in the blank with the correct name for the symbol

1.  WH — **WATTHOUR METER**

2. 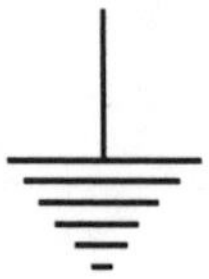 **GROUND**

3.  **CIRCUIT BREAKER**

4. 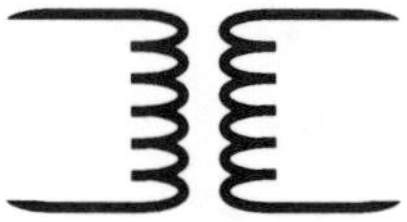 **TRANSFORMER**

5.  **FUSIBLE ELEMENT**

6.  **CONTROLLER**

7. S_F — **SWITCH FUSED**

8. 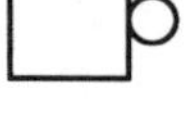 **BELL**

9.  M — **MOTOR**

10.  **BUZZER**

JOURNEYMAN WIREMAN QUIZ #2
SWITCH QUESTIONS

• *Circle the correct answer.*

1. Which of the following is properly connected?

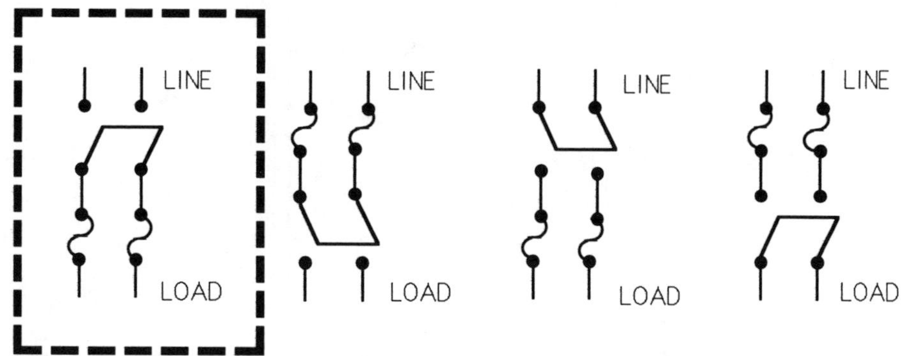

2. Which of the following double-pole double-throw switches is properly connected as a reversing switch?

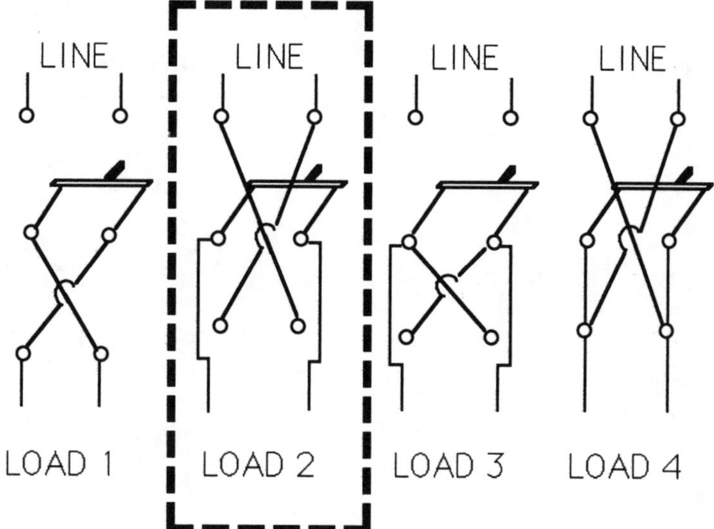

JOURNEYMAN WIREMAN QUIZ #3
TOOL IDENTIFICATION QUESTIONS

• *Fill in the blank with the correct name for the tool.*

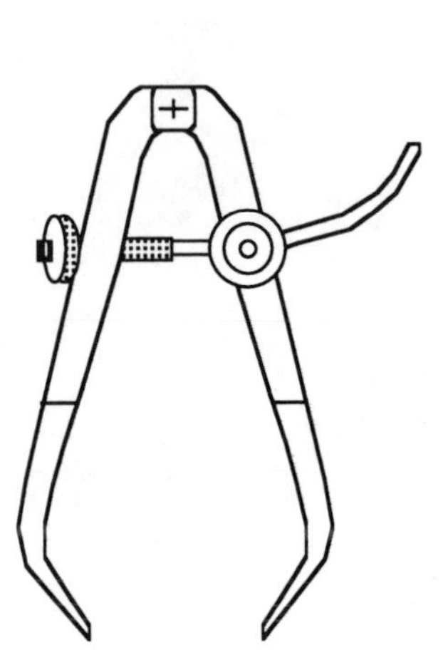

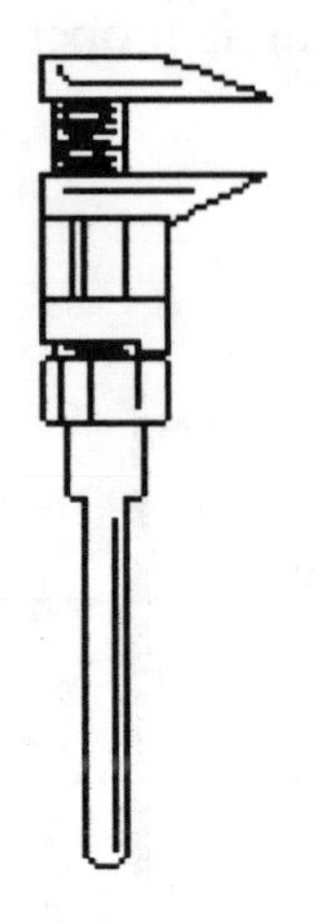

1. **Caliper**
2. **Phillips-head Screwdriver**
3. **Monkey wrench**
4. **Wood Chisel**

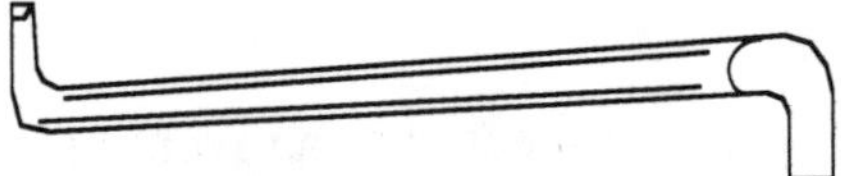

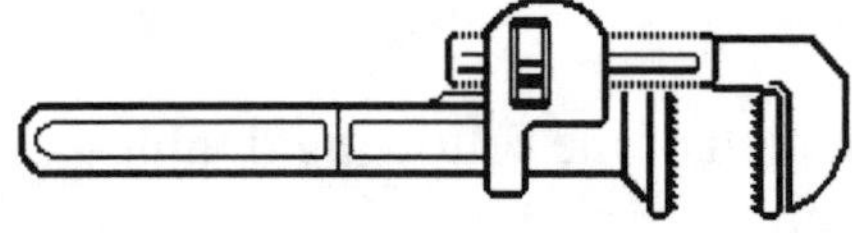

5. **Offset scewdriver**
6. **Pipe wrench**

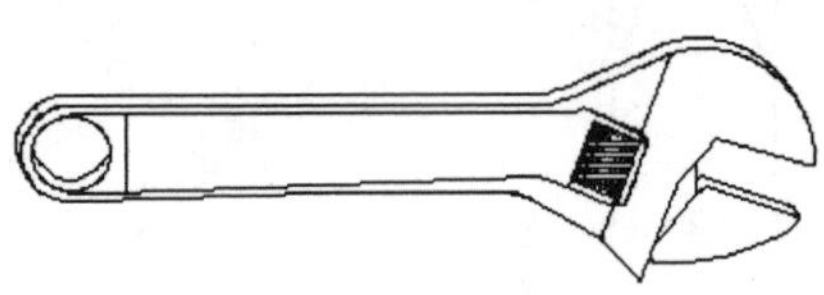

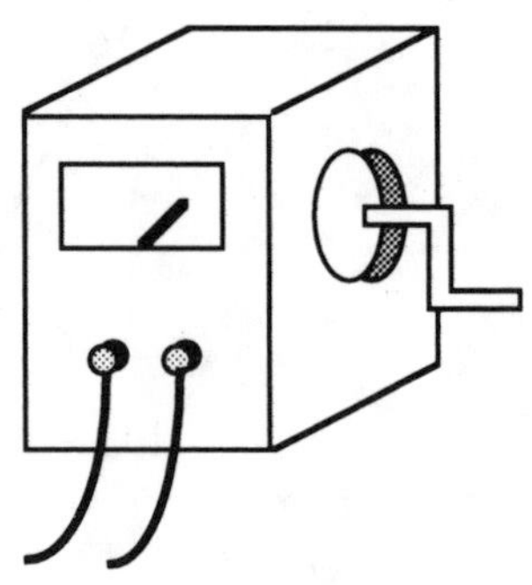

7. **ADJUSTABLE WRENCH**

8. **MEGGER**

JOURNEYMAN WIREMAN QUIZ #4
WIRING METHODS

• *Fill in the blank with the correct name for the wiring method or equipment.*

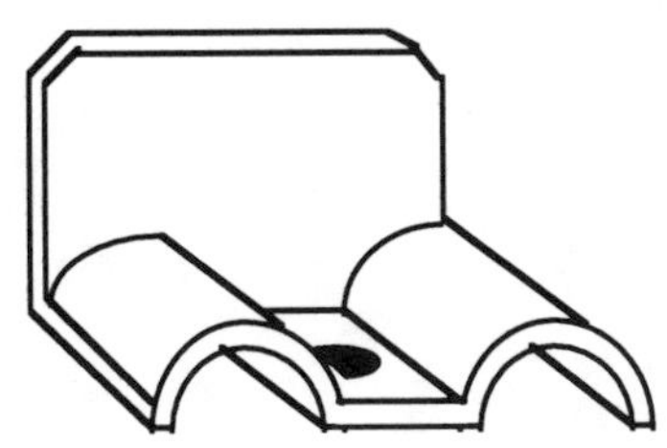

1. CABLE CLAMP

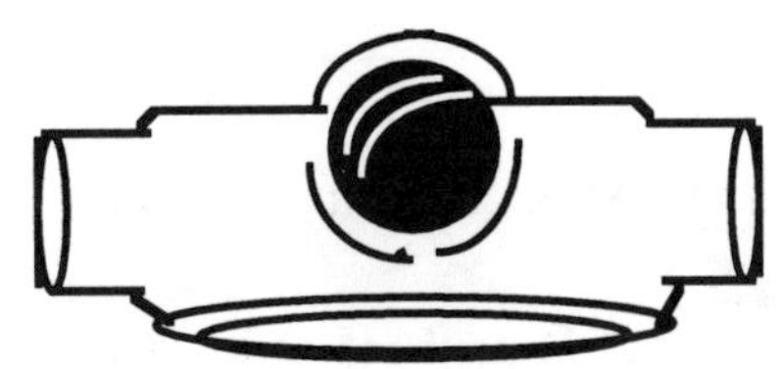

2. CONDUIT TEE

3. TRI-PLEX RECEPTACLE

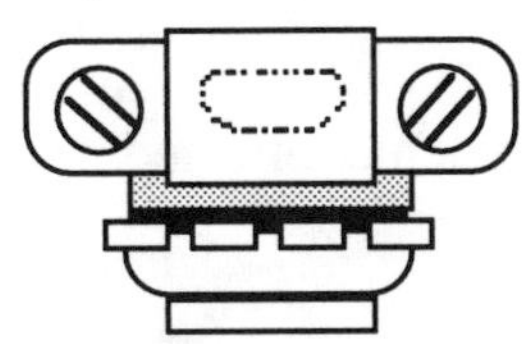

4. TWO-SCREW CONNECTOR

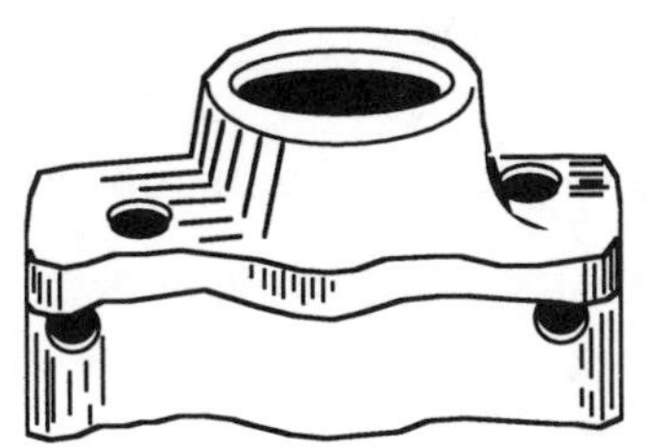

5. OPEN-WIRING RECEPTACLE

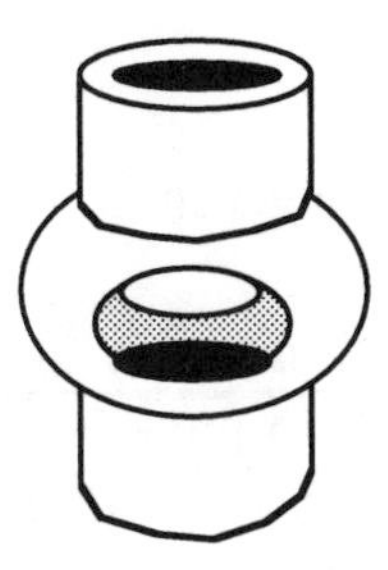

6. HICKEY

6. FIXTURE STUD

7. ARMORED CABLE

JOURNEYMAN WIREMAN QUIZ #5
PRACTICAL QUESTIONS

•*Circle the correct answer:*

1. A switch is a device for ____.

I. making or breaking connections
II. changing connections
III. interruption of circuit under short-circuit conditions

(a) I only **(b) I and II only** **(c) II and III only** **(d) I, II and III**

2. One of the essential functions of any switch is to maintain a ____.

(a) good high-resistance contact in the closed position
(b) good low-resistance contact in the closed position
(c) good low-resistance contact in the open position
(d) none of these

3.For a given line voltage, four heater coils will consume the most power when connected ____.

(a) all in series
(b) all in parallel
(c) with 2 parallel pairs in series
(d) 1 pair in parallel with the other two in series

4. All edges that are invisible should be represented in a drawing by lines that are ____.

(a) dotted
(b) broken
(c) curved
(d) solid

5.Two switches in one box under one face-plate is called a ____.

(a) double-pole switch
(b) two-gang switch
(c) 2-way switch
(d) 4-way switch

JOURNEYMAN WIREMAN QUIZ #6
BLUEPRINT SYMBOLS

•Fill in the blank with the correct name for the symbol

1. (symbol) R — **RANGE RECEPTACLE**

2. (symbol) — **POWER PANEL**

3. (symbol) — **SAFETY SWITCH**

4. (symbol) — **DUPLEX RECEPTACLE OUTLET**

5. D — **ELECTRIC DOOR OPENER**

6. (symbol) — **EXIT LIGHT**

7. H — **HUMIDISTAT**

8. S_3 — **SWITCH 3-WAY**

9. (symbol) — **ONE CELL**

10. J — **JUNCTION BOX**

JOURNEYMAN WIREMAN QUIZ #7
TOOL IDENTIFICATION QUESTIONS

• *Fill in the blank with the correct name for the tool.*

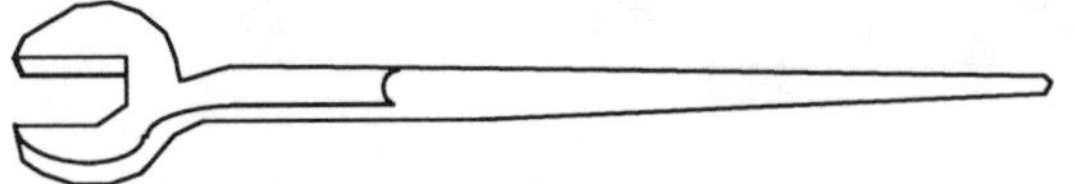

1. **ERECTION WRENCH**

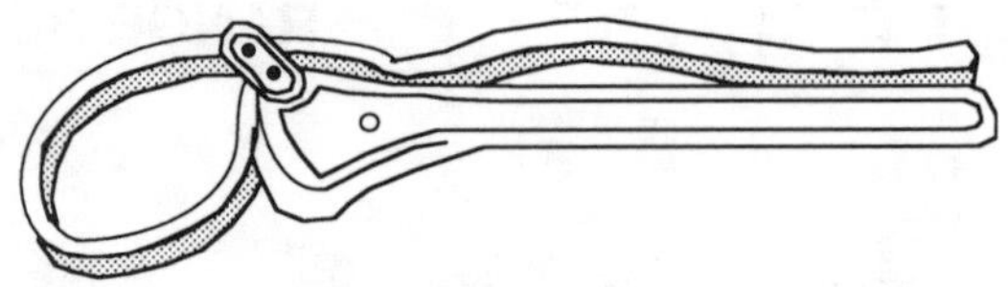
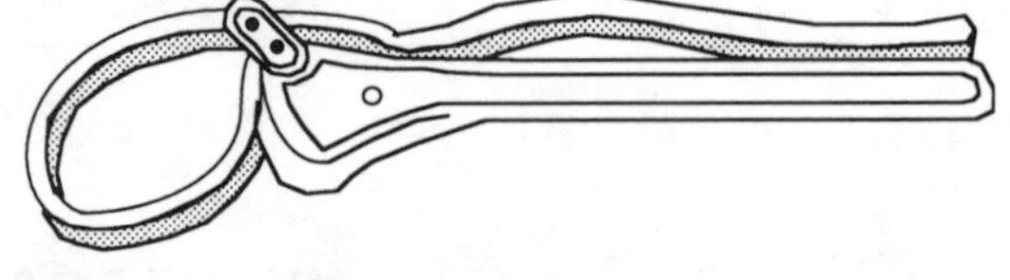

2. **STRAP WRENCH**

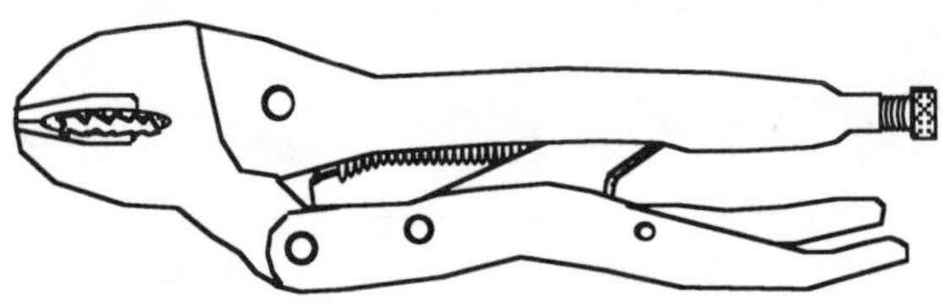

3. **LOCKING PLIERS**

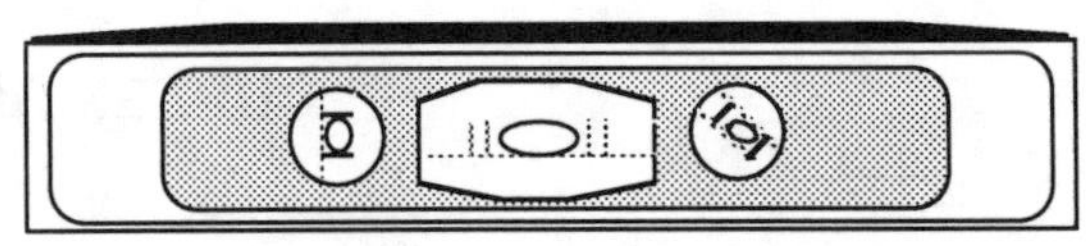

4. **TORPEDO LEVEL**

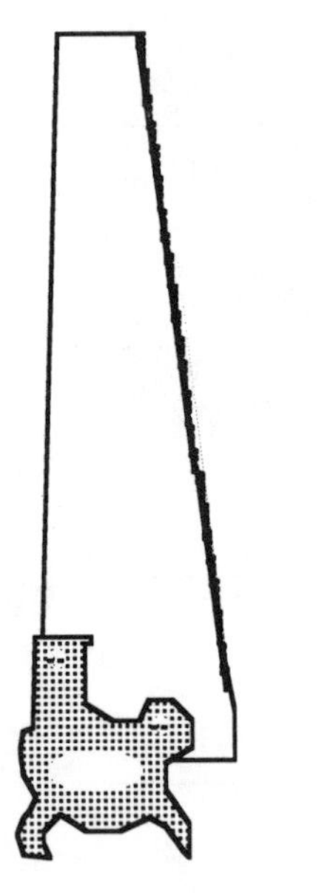

5. **HAND SAW**

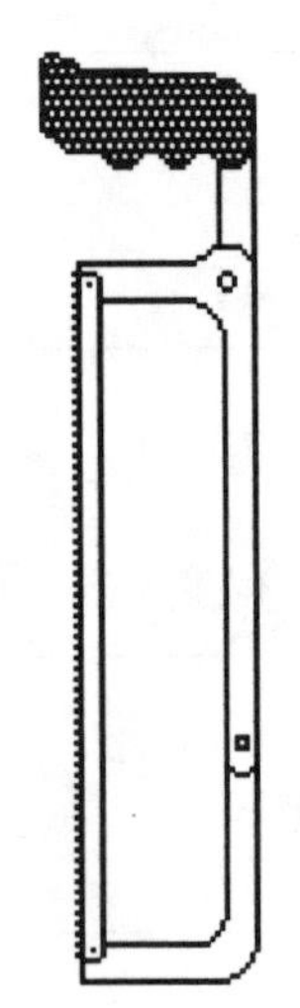

6. **HACK SAW**

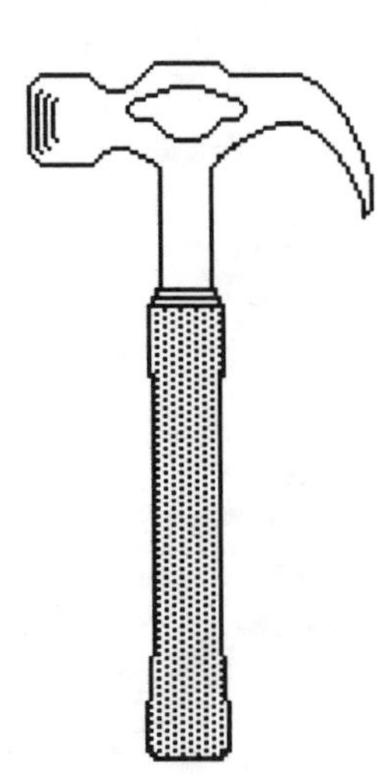

7. **CLAW HAMMER**

JOURNEYMAN WIREMAN QUIZ #8
CONNECTIONS

1. Connect the following three single-phase transformers delta-wye three-phase.

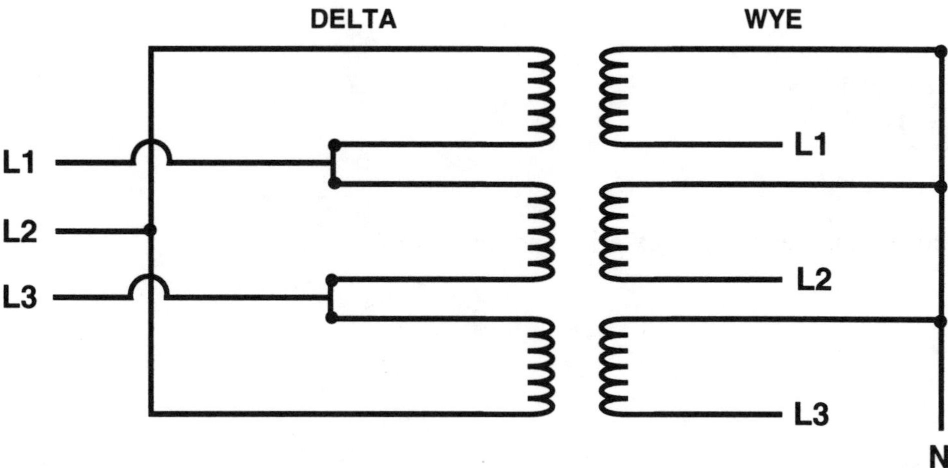

2. Which of the following is the correct wiring to a light controlled by two 3-way switches?

• *Circle the correct diagram.*

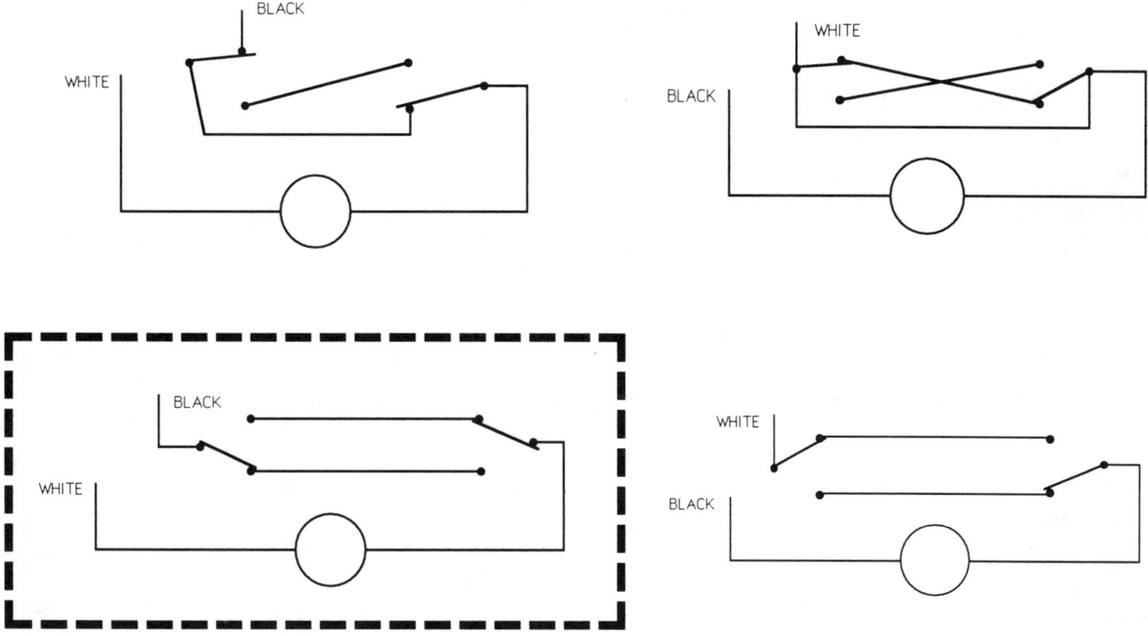

JOURNEYMAN WIREMAN QUIZ #9
PRACTICAL QUESTIONS

•*Circle the correct answer:*

1. When a gauge number such as "No.4" is used in connection with a twist drill, it refers to the ____.

(a) length
(b) hardness
(c) number of cutting edges
(d) diameter

2. Galvanized conduit has a finish exterior and interior of ____.

(a) lead
(b) copper
(c) zinc
(d) nickel

3. When stripping insulation from an aluminum conductor ____.

I. remove insulation as you would sharpen a pencil
II. ring the conductor and slip the insulation off the conductor
III. peel the insulation back and then cut outwards

(a) I, II and III (b) I and II only (c) I and III only (d) II and III only

4. A ____ is used to test the electrolyte of a battery.

(a) growler
(b) hydrometer
(c) manometer
(d) voltmeter

5. A drawing showing the floor arrangement of a building is referred to as a(an) ____.

(a) perspective
(b) isometric
(c) surface G.B.
(d) plan

JOURNEYMAN WIREMAN QUIZ #10
CONTROLS

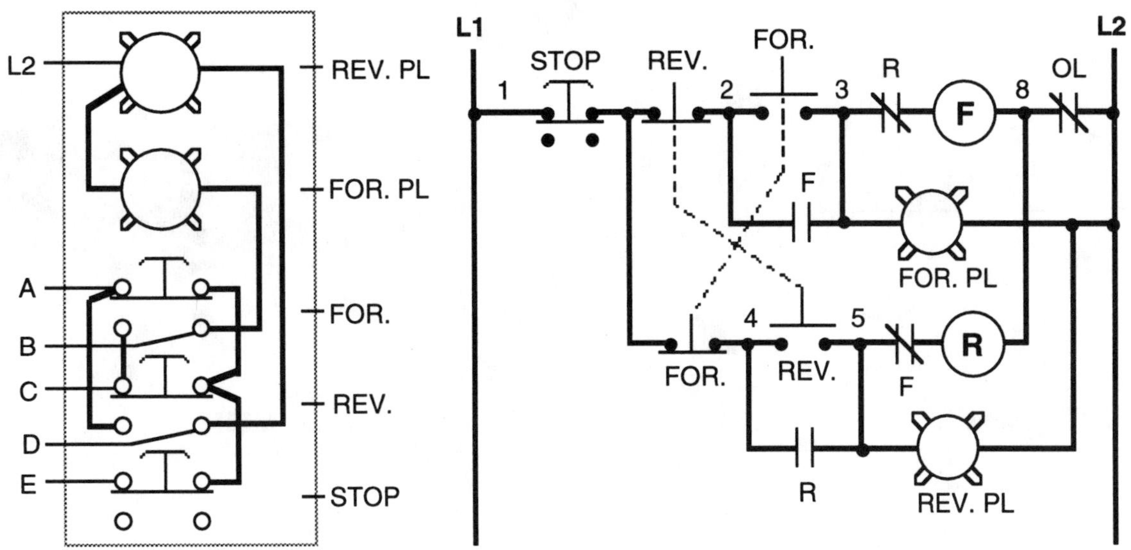

1. Conductor "A" on the push button station is conductor number ____ shown in the control circuit schematic.

(a) 1 (b) 2 (c) 3 **(d) 4** (e) 5

2. Conductor "B" on the push button station is conductor number ____ shown in the control circuit schematic.

(a) 1 (b) 2 **(c) 3** (d) 4 (e) 5

3. Conductor "C" on the push button station is conductor number ____ shown in the control circuit schematic.

(a) 1 **(b) 2** (c) 3 (d) 4 (e) 5

4. Conductor "D" on the push button station is conductor number ____ shown in the control circuit schematic.

(a) 1 (b) 2 (c) 3 (d) 4 **(e) 5**

5. Conductor "E" on the push button station is conductor number ____ shown in the control circuit schematic.

(a) 1 (b) 2 (c) 3 (d) 4 (e) 5

JOURNEYMAN WIREMAN QUIZ #11
TOOL IDENTIFICATION QUESTIONS

• *Fill in the blank with the correct name for the tool.*

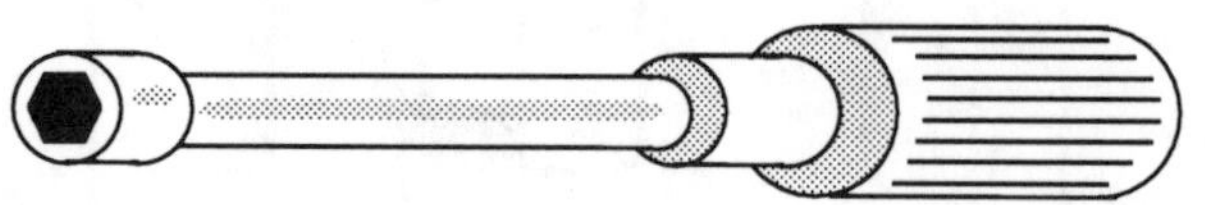

1. NUT DRIVER

2. AMP CLAMP METER

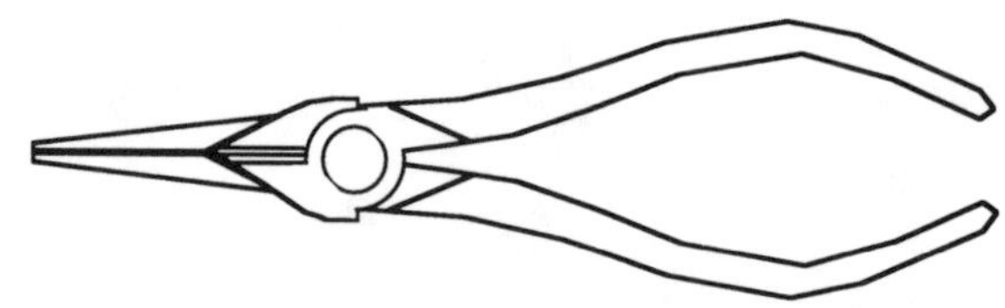

3. LONG NOSE PLIERS

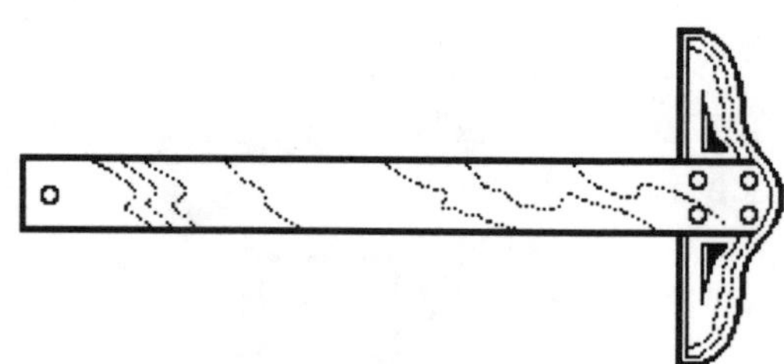

4. T - SQUARE

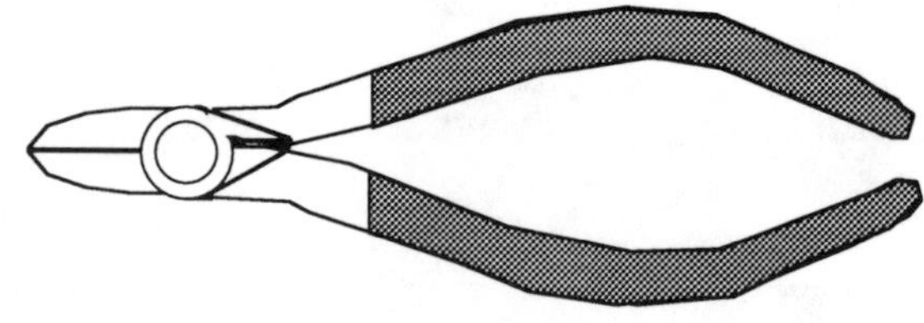

5. DIAGONAL CUTTING PLIER

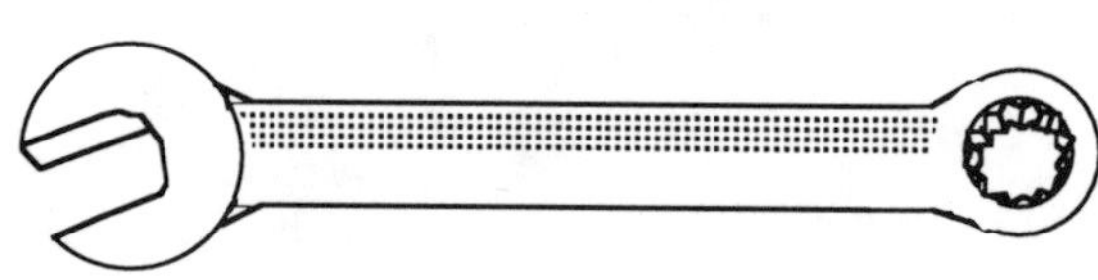

6. COMBINATION WRENCH

JOURNEYMAN WIREMAN QUIZ #12
BLUEPRINT SYMBOLS

•Fill in the blank with the correct name for the symbol

1. S_K SWITCH KEY OPERATED
2. [solid black rectangle] BRANCH CIRCUIT LIGHTING PANEL
3. F — FIRE ALARM BELL
4. TP — TRANSFORMER PAD
5. R — FIXTURE RECESSED
6. TELEPHONE
7. T — THERMOSTAT
8. PUSHBUTTON
9. WP — WEATHERPROOF OUTLET
10. J — JUNCTION BOX

JOURNEYMAN WIREMAN QUIZ #13
PRACTICAL QUESTIONS

•*Circle the correct answer:*

1. The purpose of a clip clamp is to ____.

I. ensure good contact between the fuse terminals of cartridge fuses and the fuse clips
II. make it possible to use cartridge fuses of a smaller size than that for which the fuse clips are intended
III. prevent the accidental removal of the fuse due to vibration

(a) I, II and III (b) I only (c) II only (d) I and II only

2. An electric bell outfit would be used to check for ____.

(a) voltage
(b) ampacity
(c) continuity
(d) current

3. A hickey is ____.

(a) a tool used to bend small sizes of rigid conduit
(b) a part of a conduit
(c) not used in the electrical trade
(d) used only by a plumber

4. When using a #12-2 with ground cable, the ground ____ carry current under normal operation.

(a) will
(b) will not
(c) will sometimes
(d) none of these

5. The load side is usually wired to the blades of a knife switch to ____.

(a) prevent blowing the fuse when opening the switch
(b) make the blades dead when the switch is opened
(c) prevent arcing when the switch is opened
(d) allow changing of fuses without opening the switch

JOURNEYMAN WIREMAN QUIZ #14
WIRING METHODS

• *Fill in the blank with the correct name for the wiring method or equipment.*

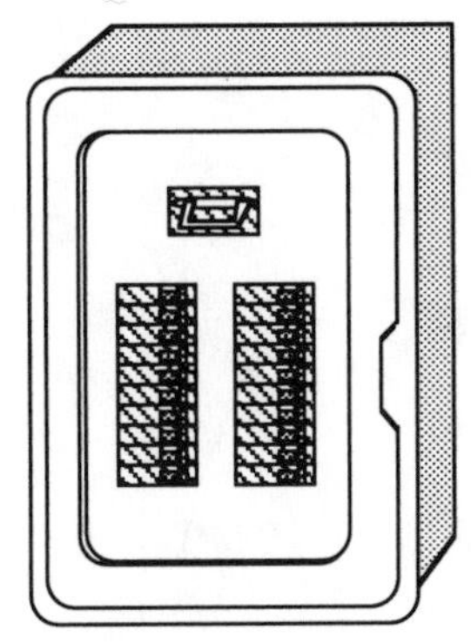

1. SINGLE RECEPTACLE
2. PANELBOARD
3. 

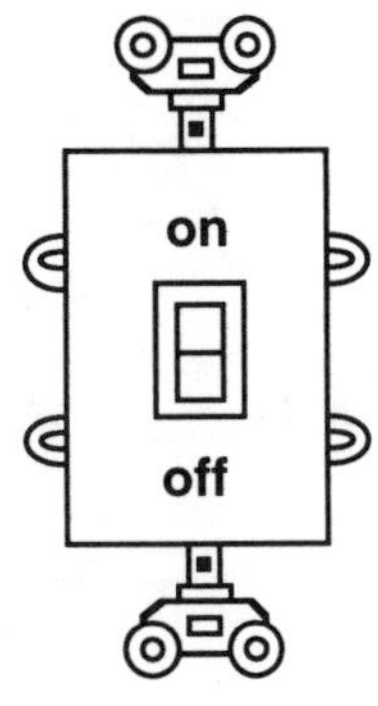

4. DOUBLE - POLE SWITCH
5. PLUGMOLD

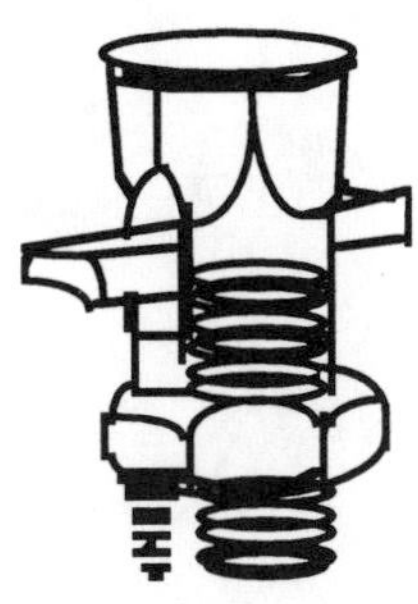

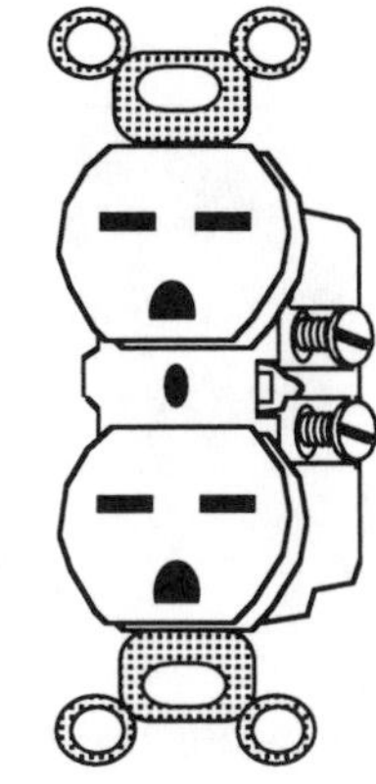

6. 

7. SPLIT-BOLT CONNECTOR
8. 240v Receptacle

JOURNEYMAN WIREMAN QUIZ #15
TOOL IDENTIFICATION QUESTIONS

• *Fill in the blank with the correct name for the tool.*

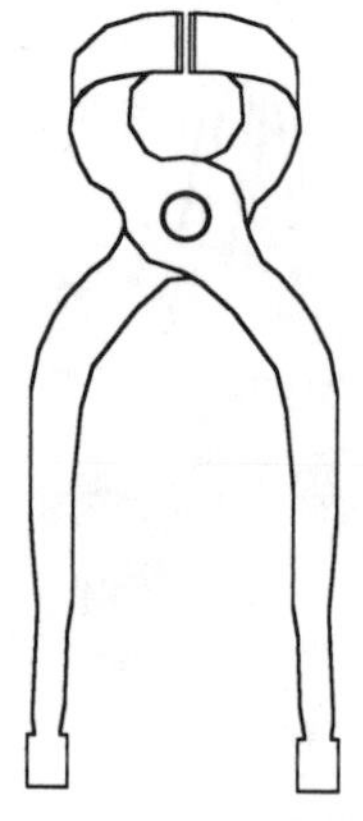

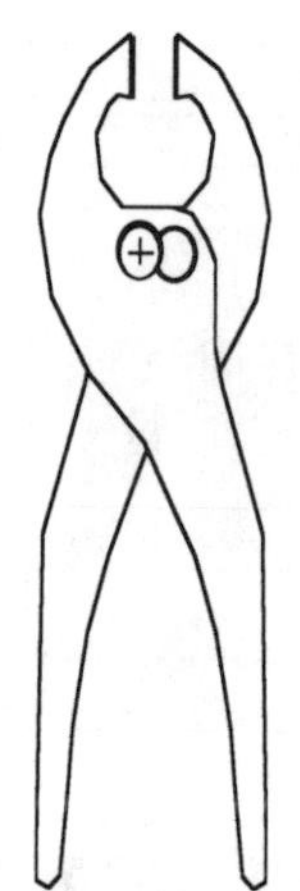

1. END CUTTING PLIERS

2. SLIP-JOINT PLIERS

3. BLOW TORCH

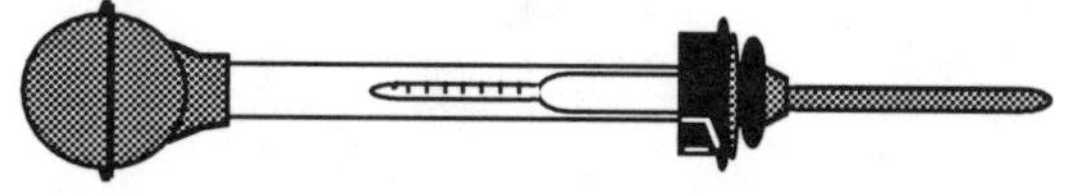

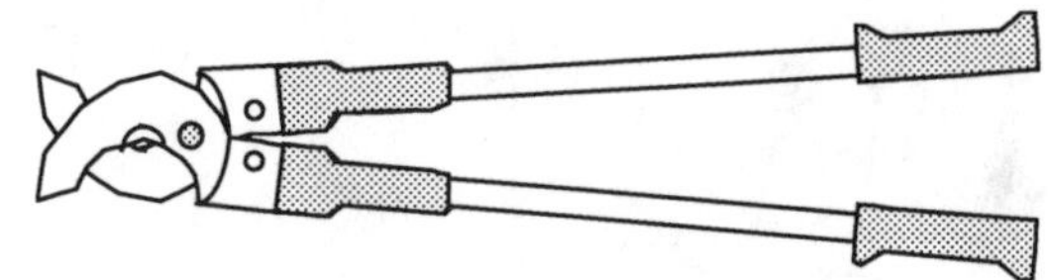

4. HYDROMETER

5. CABLE CUTTERS

6. WIRE GAUGE

7. MICROMETER

JOURNEYMAN WIREMAN QUIZ #16
BLUEPRINT SYMBOLS

•Fill in the blank with the correct name for the symbol

1. S_P — SWITCH WITH PILOT

2. SINGLE SPECIAL-PURPOSE RECEPTACLE OUTLET

3. (F) — FAN OUTLET

4. F — FIRE ALARM HORN

5. DUPLEX OUTLET, SPLIT CIRCUIT

6. SINGLE BRANCH CIRCUIT HOME RUN TO PANEL (3-wire)

7. CEILING OUTLET

8. TV — TELEVISION OUTLET

FLUORESCENT FIXTURE

10. — — — WIRING CONCEALED IN FLOOR

JOURNEYMAN WIREMAN QUIZ #17
WIRING METHODS

• *Fill in the blank with the correct name for the wiring method or equipment.*

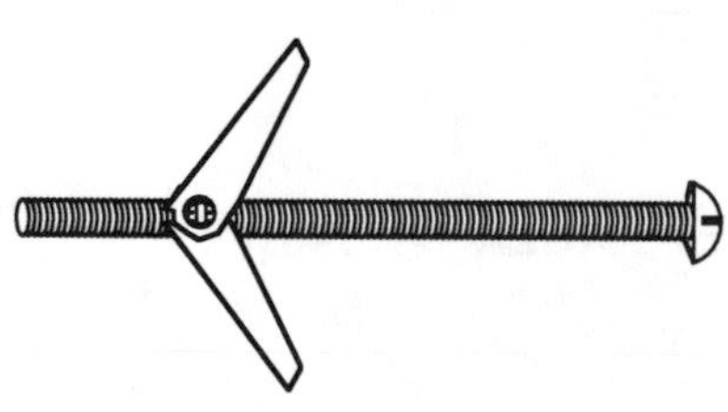

1. **TOGGLE BOLT**

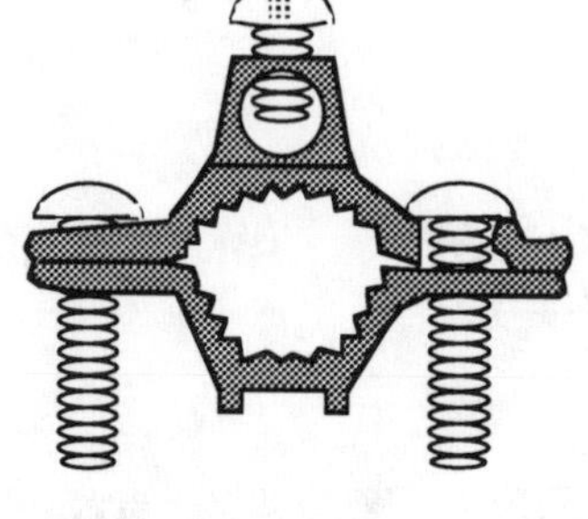

2. **GROUND CLAMP**

3. **NON-METALLIC SHEATHED CABLE**

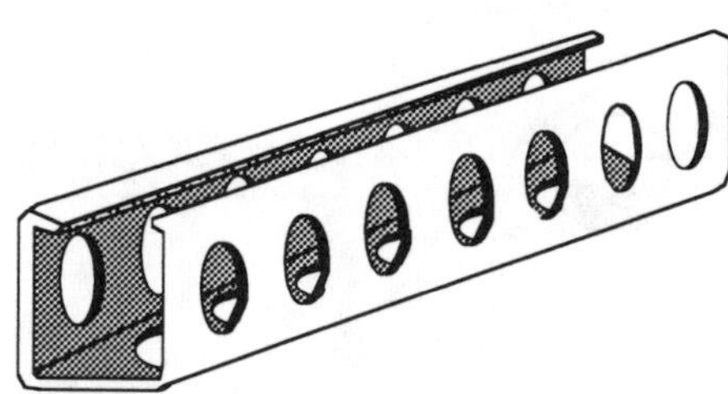

4. **KINDORF CHANNEL**

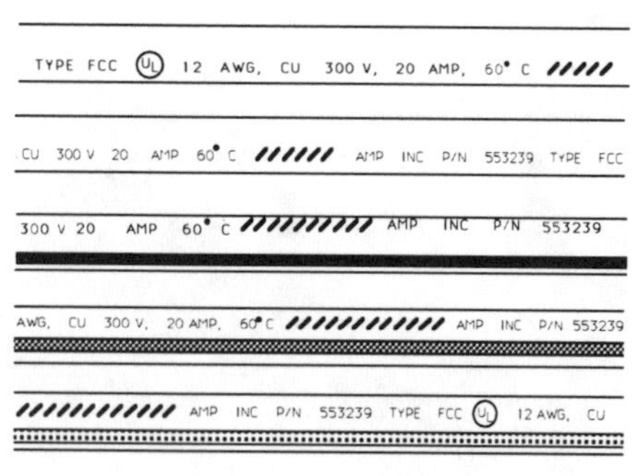

5. **FLAT CONDUCTOR CABLE**

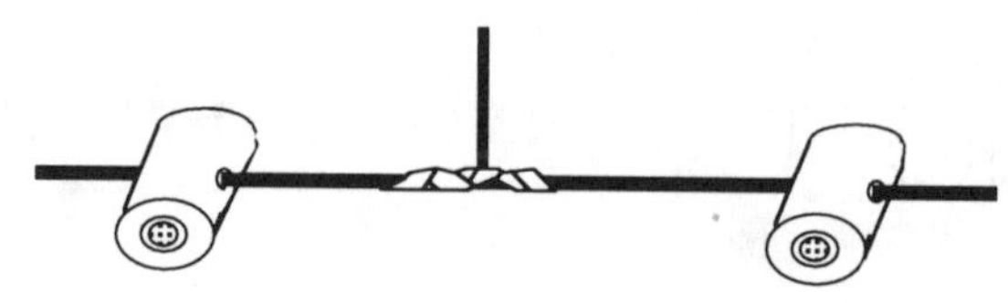

6. **OPEN WIRING**

JOURNEYMAN WIREMAN QUIZ #18
PRACTICAL QUESTIONS

•*Circle the correct answer:*

1. What is the function of a neon glow tester?

I. determines if circuit is alive
II. determines polarity of DC circuits
III. determines if circuit is AC or DC current

(a) I only (b) II only (c) III only (d) I, II and III

2. A conduit coupling is sometimes tightened by using a strap wrench rather than a Stillson wrench. The strap wrench is used when it is important to avoid ____.

(a) crushing the conduit
(b) bending the conduit
(c) stripping the threads
(d) damaging the outside finish

3. With respect to a common light bulb, it is correct to state that the ____.

(a) circuit voltage has no effect on the life of the bulb
(b) base has a left hand thread
(c) filament is made of carbon
(d) lower wattage bulb has the higher resistance

4. A multimeter is a combination of ____.

(a) ammeter, ohmmeter and wattmeter
(b) voltmeter, ohmmeter and ammeter
(c) voltmeter, ammeter and megger
(d) voltmeter, wattmeter and ammeter

5. Two 120 volt light bulbs connected in series across 240 volt will ____.

(a) burn at full brightness
(b) burn at half-brightness
(c) burn out quickly
(d) flicker with the cycle

JOURNEYMAN WIREMAN QUIZ #19
TOOL IDENTIFICATION QUESTIONS

• *Fill in the blank with the correct name for the tool.*

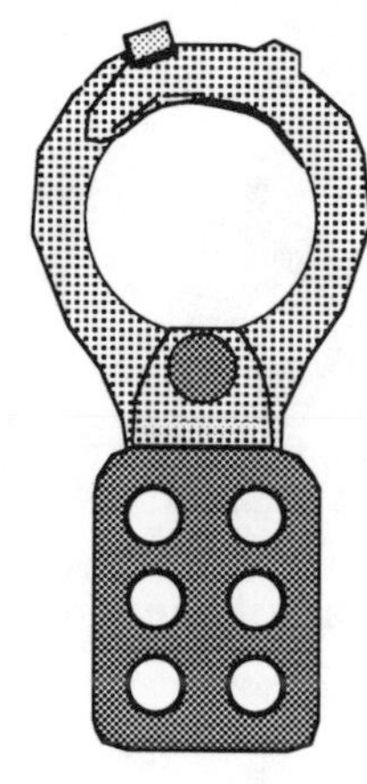

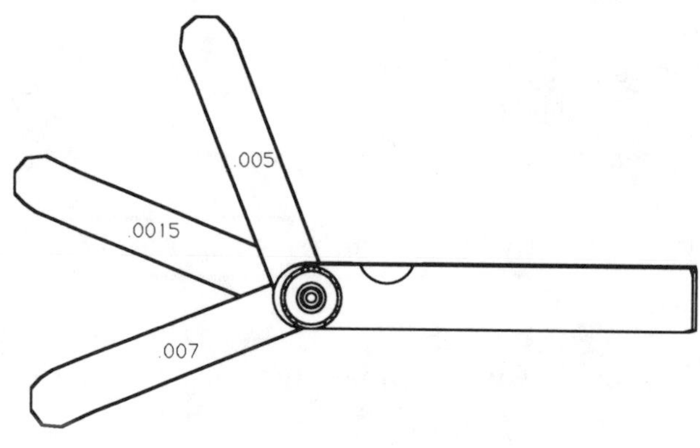

1. LOCKOUT

2. FEELER GAUGE

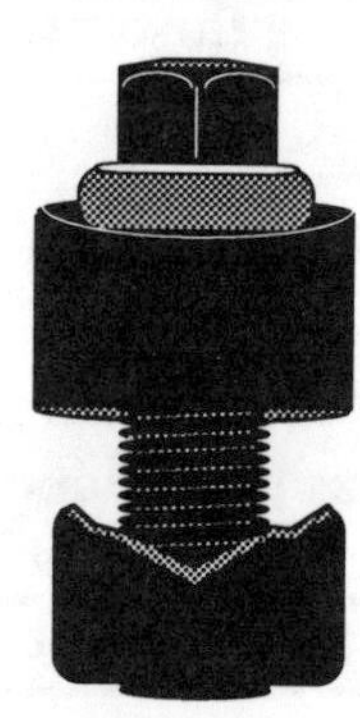

3. UTILITY KNIFE

4. KNOCKOUT

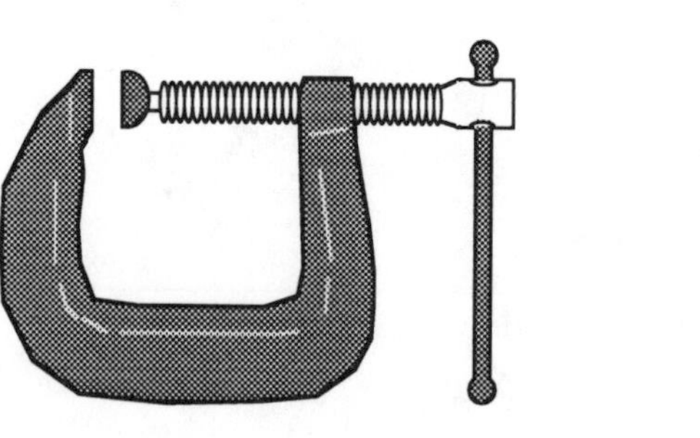

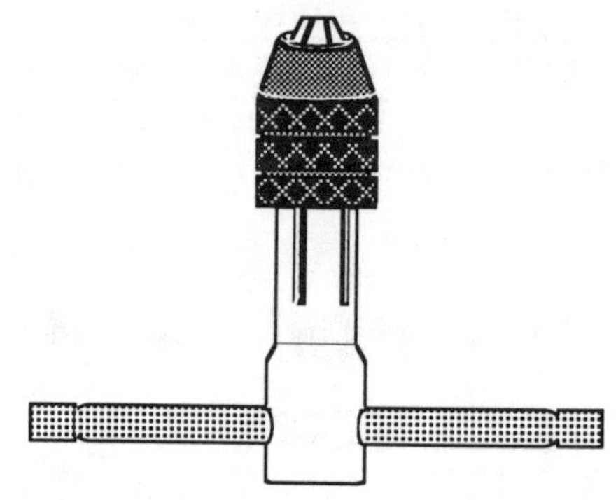

5. "C" CLAMP

6. TAP HANDLE

JOURNEYMAN WIREMAN QUIZ #20
BLUEPRINT SYMBOLS

•Fill in the blank with the correct name for the symbol

1. (B) BLANKED OUTLET

2. BARE-LAMP FLUORESCENT STRIP

3. SINGLE BRANCH CIRCUIT HOME RUN TO PANEL

4. HEATING PANEL

5. S_T TIME SWITCH

6. CH CHIME

7. — — — — — WIRING EXPOSED

8. FLOOR OUTLET

9. WALL BRACKET

10. 3 TRIPLEX OUTLET

JOURNEYMAN WIREMAN QUIZ #21
CONTROLS

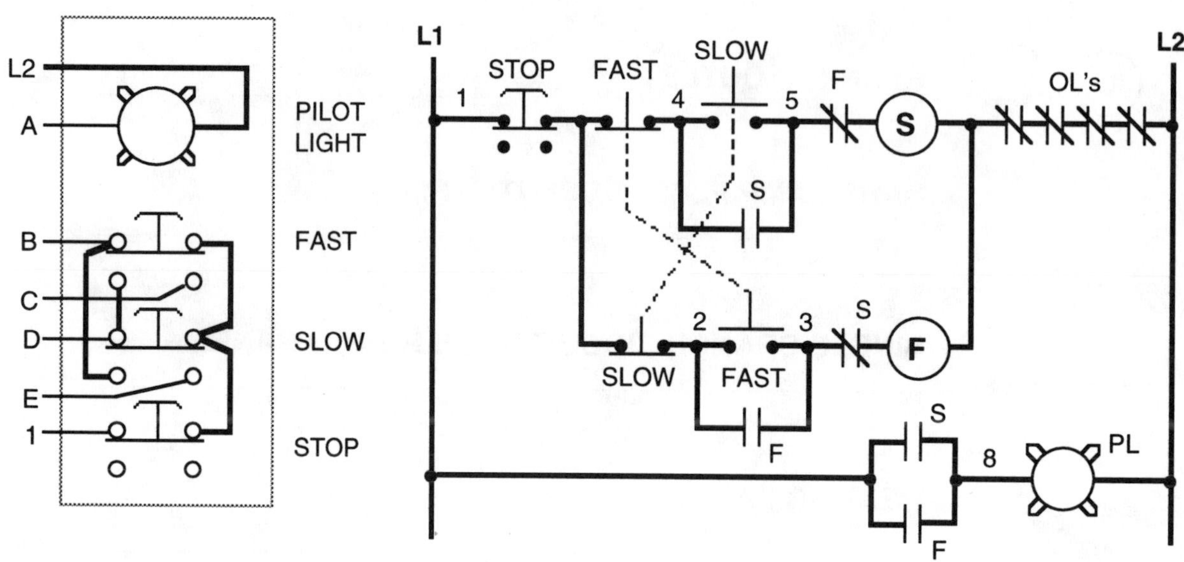

1. "**A**" is actually point ______ on the control diagram.
   a) L1 b) L2 **c) 8** d) 6

2. "**D**" is actually point ______ on the control diagram.
   a) 5 b) 4 c) 3 **d) 2**

3. "**B**" is actually point ______ on the control diagram.
   a) 4 b) 3 c) 2 d) L1

4. "**C**" is actually point ______ on the control diagram.
   a) 4 **b) 3** c) 2 d) L1

5. "**E**" is actually point ______ on the control diagram.
   a) 4 **b) 5** c) 1 d) 2

JOURNEYMAN WIREMAN QUIZ #22
TOOL IDENTIFICATION QUESTIONS

• Fill in the blank with the correct name for the tool.

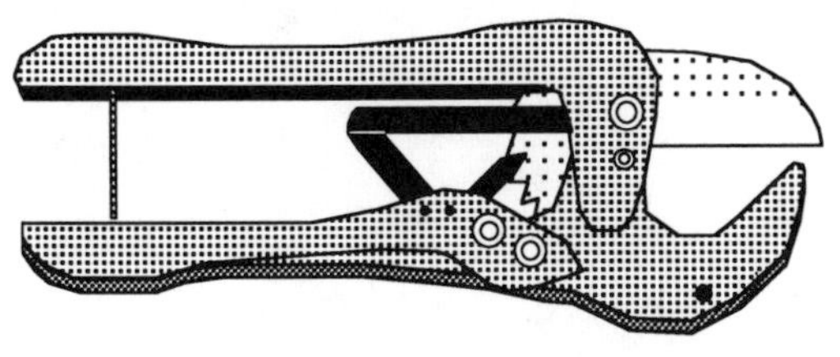

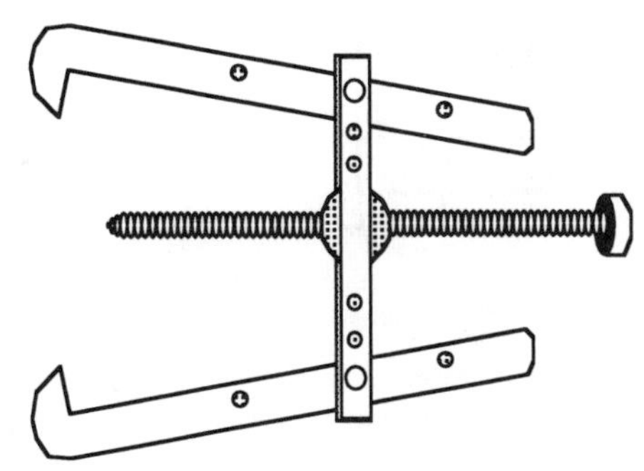

1. PVC CUTTER

2. PULLEY and GEAR PULLERS

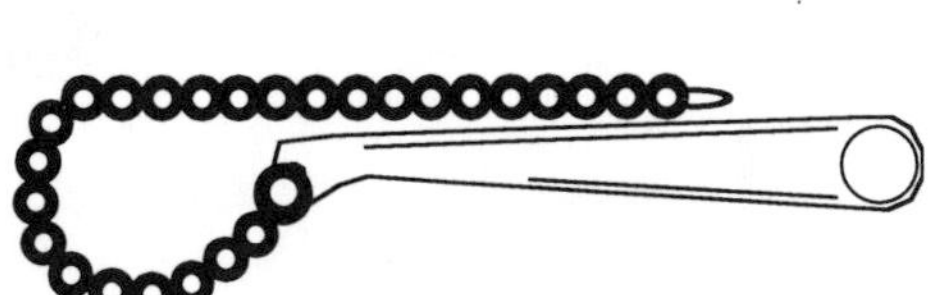

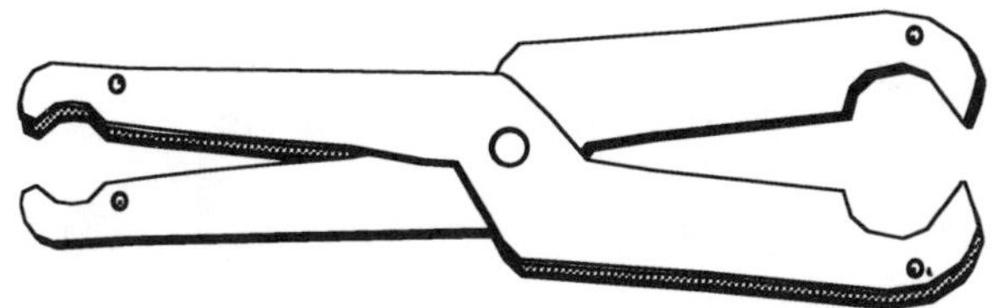

3. CHAIN WRENCH

4. FUSE PULLER

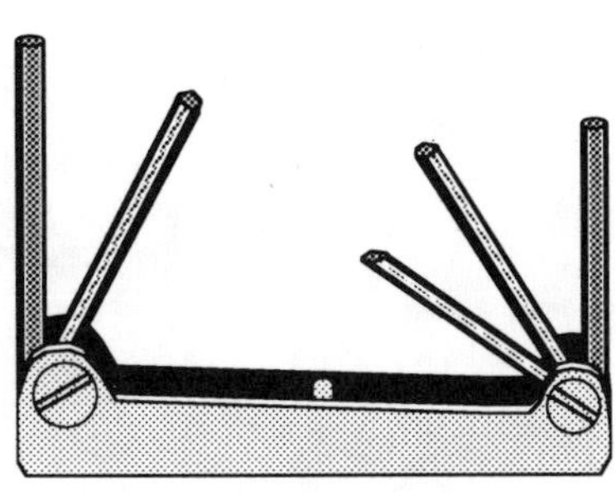

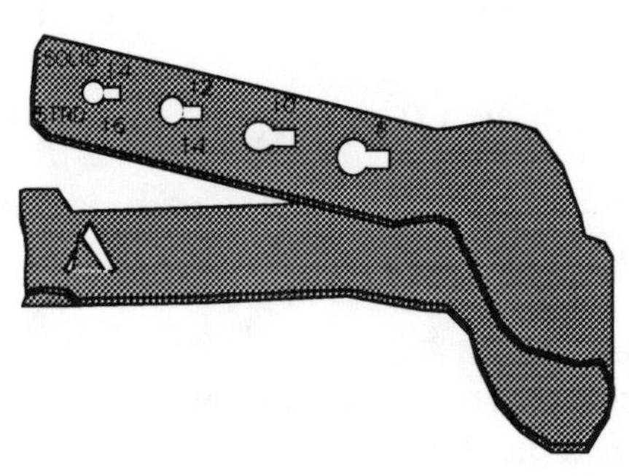

5. HEX KEY WRENCH SET

6. CABLE STRIPPER

JOURNEYMAN WIREMAN QUIZ #23
PRACTICAL QUESTIONS

•*Circle the correct answer:*

1. What type fastener would you use to mount a box to a hollow tile wall?

(a) expansion bolts
(b) wooden plugs
(c) toggle bolts
(d) bolts with backing plates

2. When a current leaves its intended path and returns to the source bypassing the load, the circuit is ____.

(a) open
(b) shorted
(c) incomplete
(d) broken

3. A clamp-on ammeter will measure ____.

(a) voltage when clamped on a single conductor
(b) current when clamped on a multiconductor cable
(c) accurately only when parallel to the cable
(d) accurately only when clamped perpendicular to a conductor

4. Which of the following statements about mounting single-throw knife switches in a vertical position is/are correct?

I. the switch shall be mounted so that the blade hinge is at the bottom
II. the supply side of the circuit shall be connected to the bottom of the switch

(a) I only (b) II only (c) both I and II (d) neither I nor II

5. A hook on the end of a fish tape is **not** to ____.

(a) keep it from catching on joints and bends
(b) tie a swab to
(c) tie the wires, to be pulled, to
(d) protect the end of the wire

JOURNEYMAN WIREMAN QUIZ #24
BLUEPRINT SYMBOLS

•Fill in the blank with the correct name for the symbol

1.  WIRING CONNECTED

2.  PHASE

3. WIRING OR CONDUIT TURNED DOWN

4. 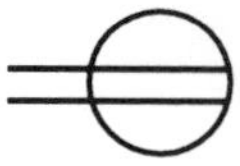 GR GROUNDED DUPLEX RECEPTACLE

5. WIRING CONCEALED IN CEILING OR WALL

6.  C CLOCK

7.  TWO BRANCH CIRCUIT HOME RUNS TO PANEL

8. 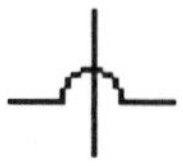 WIRING CROSSING NOT CONNECTED

9.  WIRING OR CONDUIT TURNED UP

10.  PS LAMPHOLDER WITH PULL SWITCH

JOURNEYMAN WIREMAN QUIZ #25
TOOL IDENTIFICATION QUESTIONS

• *Fill in the blank with the correct name for the tool.*

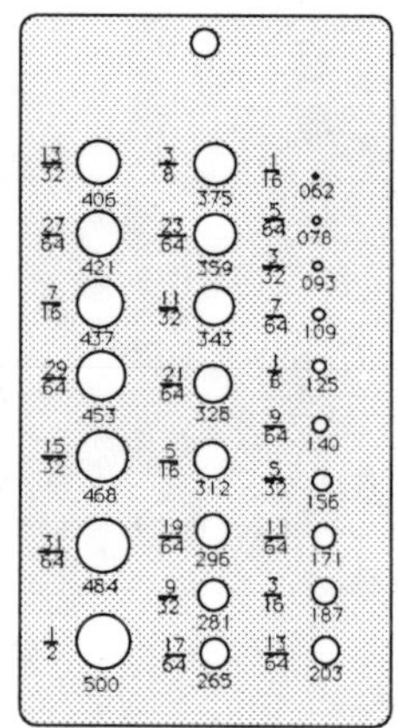

1. DRILL BIT GAUGE

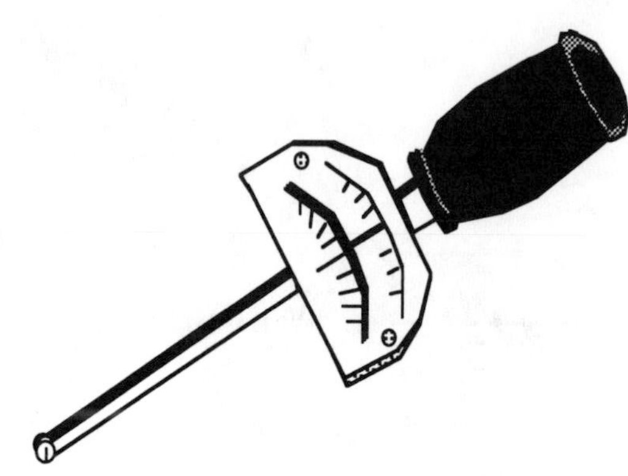

2. TORQUE WRENCH

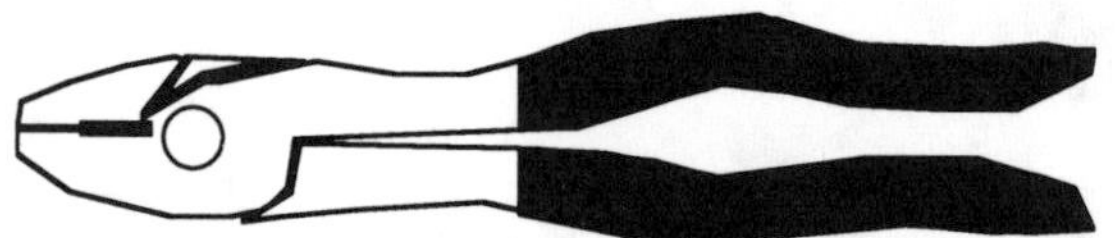

3. SIDE CUTTING PLIERS

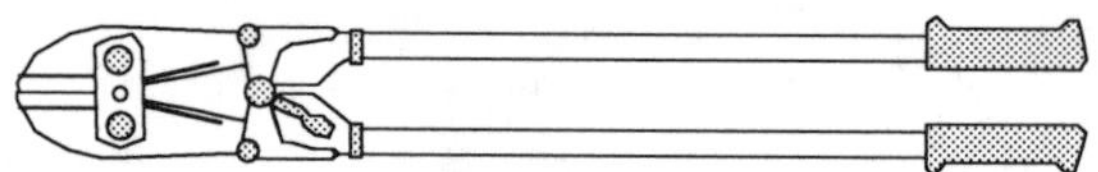

4. BOLT CUTTERS

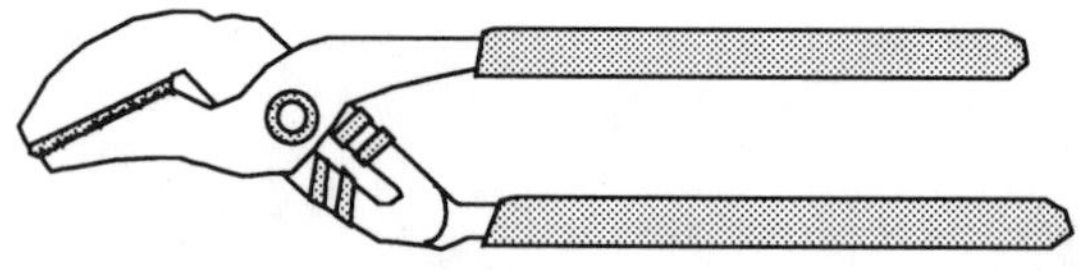

5. PUMP PLIERS

6. SCRATCH AWL

JOURNEYMAN WIREMAN QUIZ #26
WIRING METHODS

• *Fill in the blank with the correct name for the wiring method or equipment.*

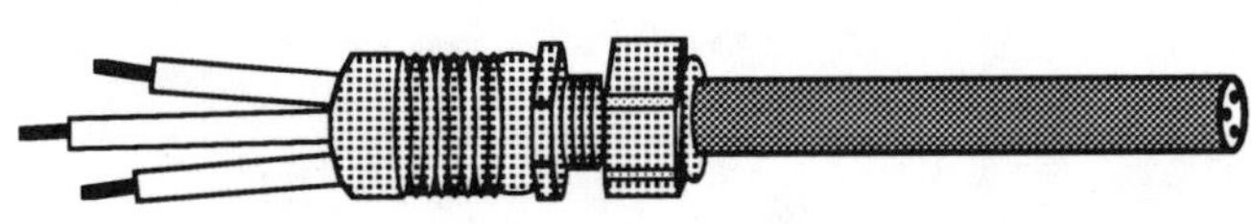

1. **MI CABLE**

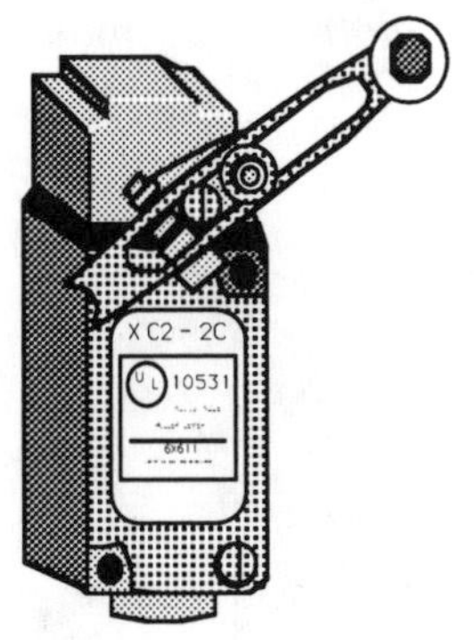

2. **LIMIT SWITCH**

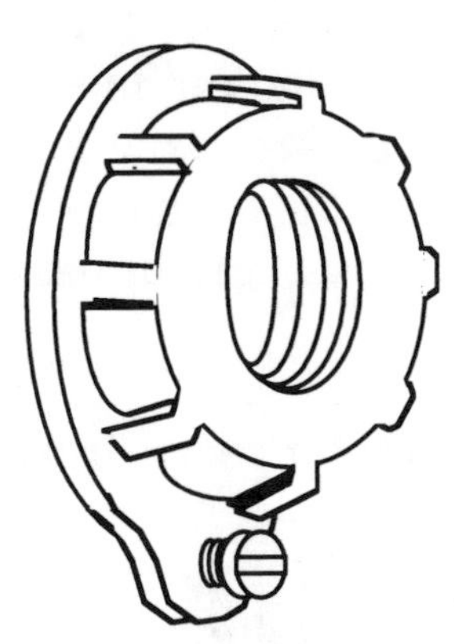

3. **GROUNDING BUSHING**

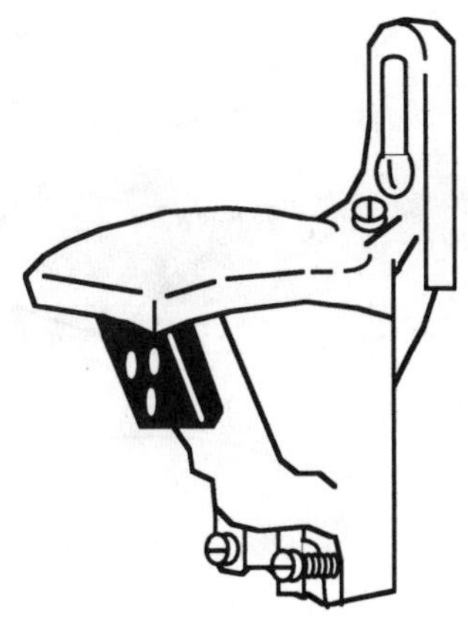

4. **SERVICE HEAD for CABLE**

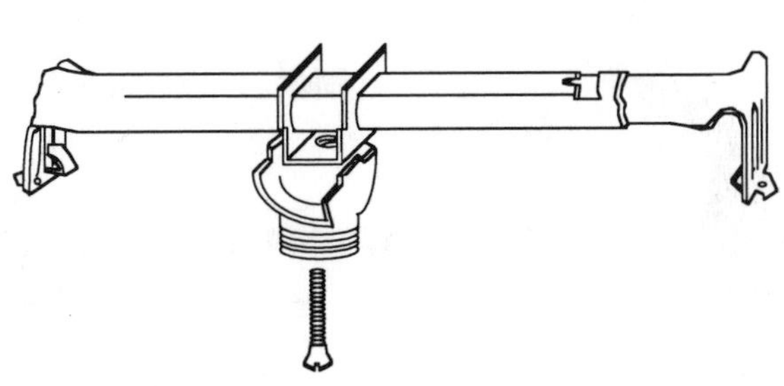

5. **BAR HANGER**

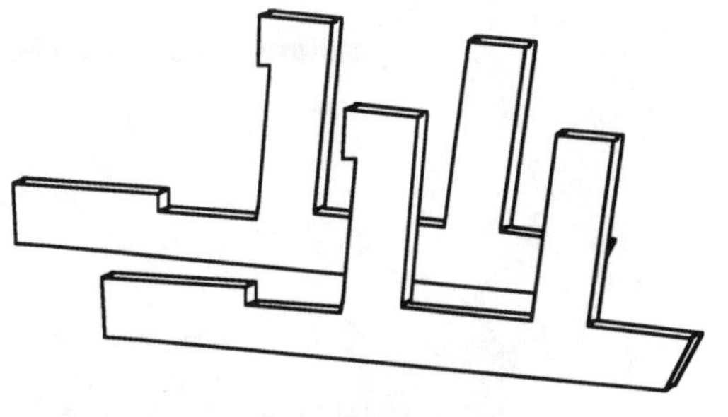

6. **MADISON STRAP**

JOURNEYMAN WIREMAN QUIZ #27

• *The top watthour meter is read at the beginning of the month, the bottom meter is read at the end of the month. How many kilowatthours were consumed?* **4435** *kwh.*

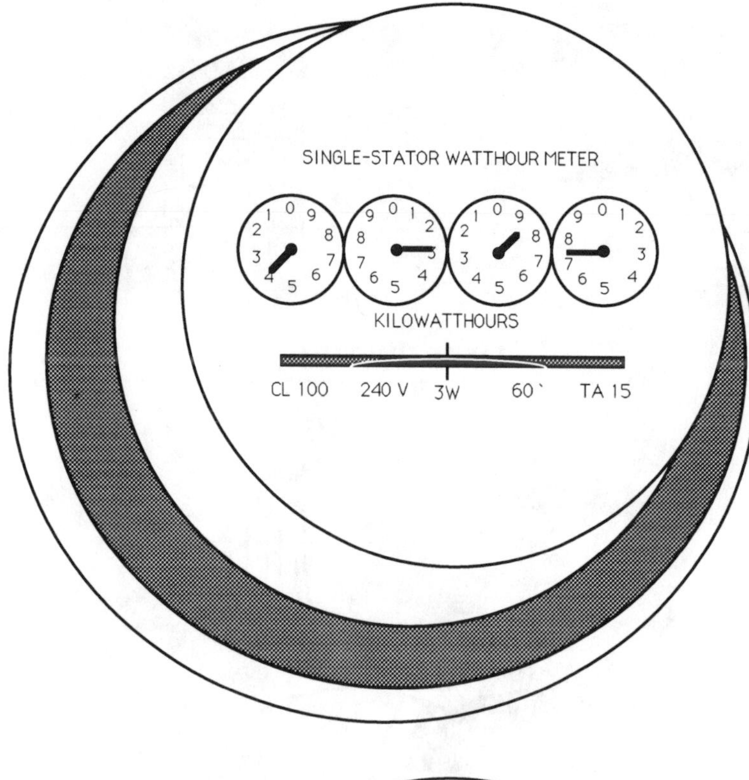

The meter is read from right to left, write down the number the pointer has passed, if the pointer has not passed a number write down the next lower number.

This meter reading from right to left -

First dial reads 7
Second dial reads 8
Third dial reads 2
Fourth dial reads 3

Meter reads 3287

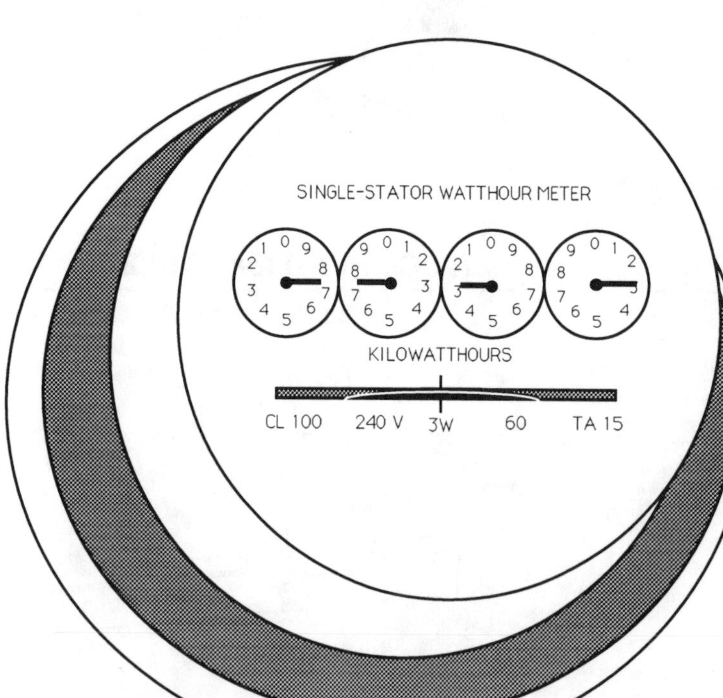

This meter reads from right to left -

First dial reads 2
Second dial reads 2
Third dial reads 7
Fourth dial reads 7

•Always write down numbers from left to right

Meter reads 7722

7722 - 3287 = 4435 kwh used

JOURNEYMAN WIREMAN QUIZ #28
PRACTICAL QUESTIONS

•*Circle the correct answer:*

1. An Erickson coupling is used to ____.

(a) join sections of EMT together
(b) connect EMT to flexible conduit
(c) to connect two sections of rigid conduit when one section cannot be turned
(d) substitute for all-thread

2. On smaller gauges of wire, they are pencil-stripped to prevent ____.

(a) hysteresis
(b) over stripping
(c) nicks in wire
(d) loosening of wire nut

3. When the term "10-32" in connection with machine bolts commonly used in lighting work, the number "32" refers to ____.

(a) bolt length
(b) bolt thickness
(c) diameter of hole
(d) threads per inch

4. A pendant light fixture is a _____ fixture.

(a) closet
(b) recessed
(c) hanging
(d) bracket

5. Since fuses are rated by amperage and voltage a fuse will work on ____.

(a) AC only
(b) AC or DC
(c) DC only
(d) any voltage

JOURNEYMAN WIREMAN QUIZ #29
CONTROLS

•*Fill in the blank below for the correct type of motor starting.*

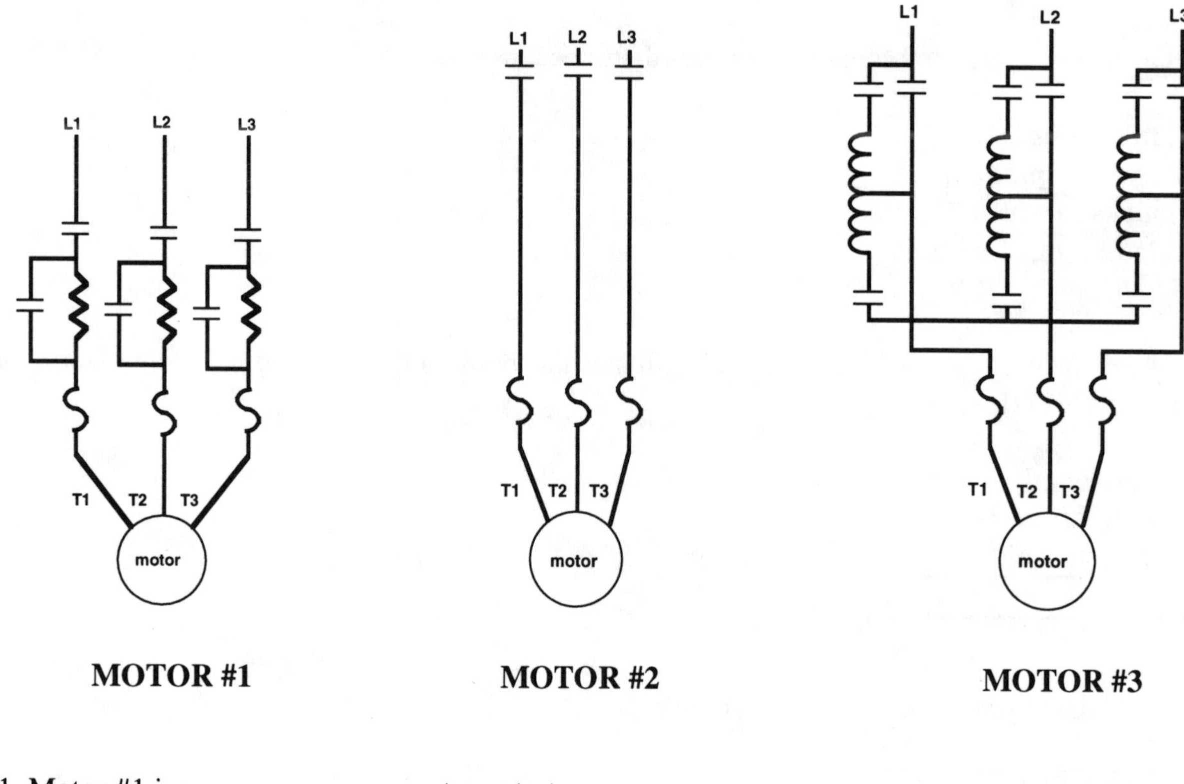

1. Motor #1 is a _______________ type start.

(a) full voltage (b) autotransformer (c) resistor

2. Motor #2 is a _______________ type start.

(a) full voltage (b) autotransformer (c) resistor

3. Motor #3 is a _______________ type start.

(a) full voltage (b) autotransformer (c) resistor

JOURNEYMAN WIREMAN QUIZ #30
TOOL IDENTIFICATION QUESTIONS

• *Fill in the blank with the correct name for the tool.*

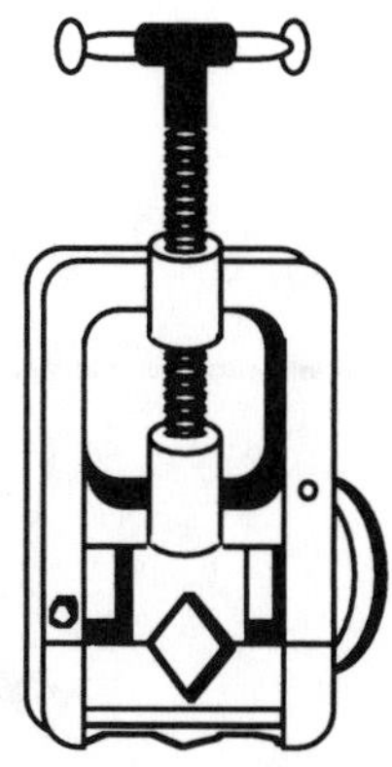

1. **Wheel & Roller pipe cutter**

2. **Pipe vise**

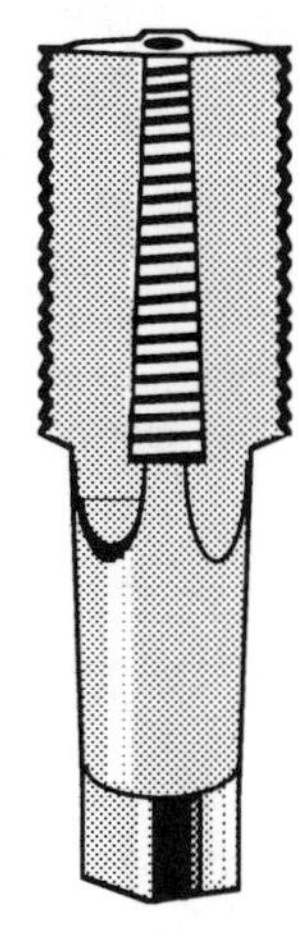

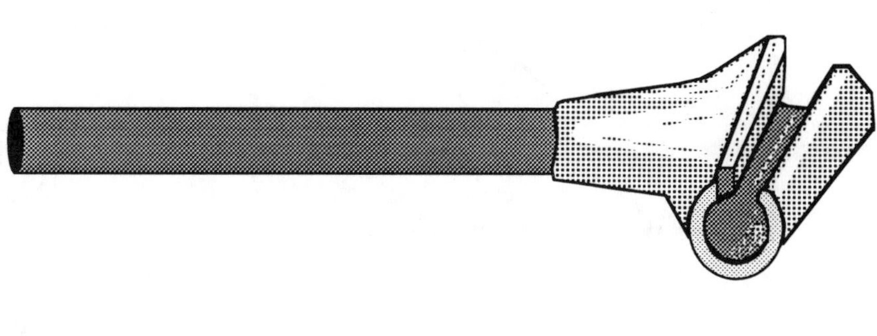

3. **Pipe thread tap**

4. **Hickey conduit bender**

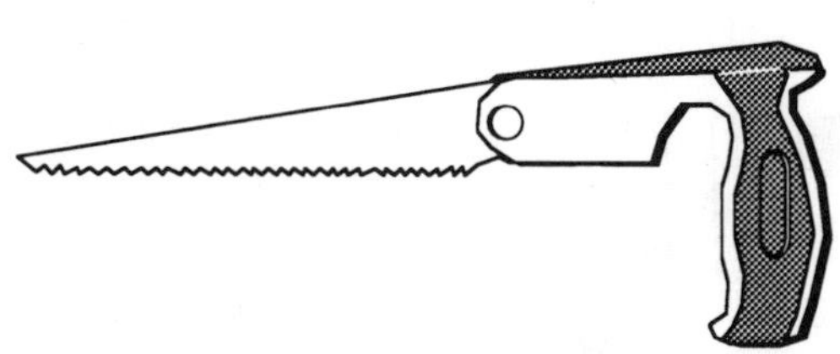

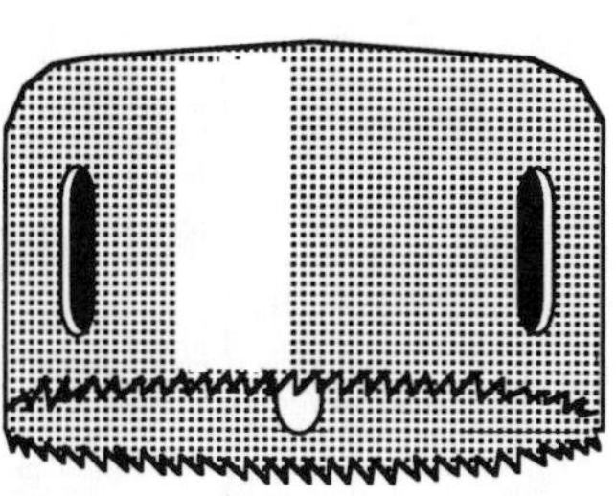

5. **Compass saw**

6. **Hole saw**

JOURNEYMAN WIREMAN QUIZ #31
INSTALLATION

•Circle the correct installation method.

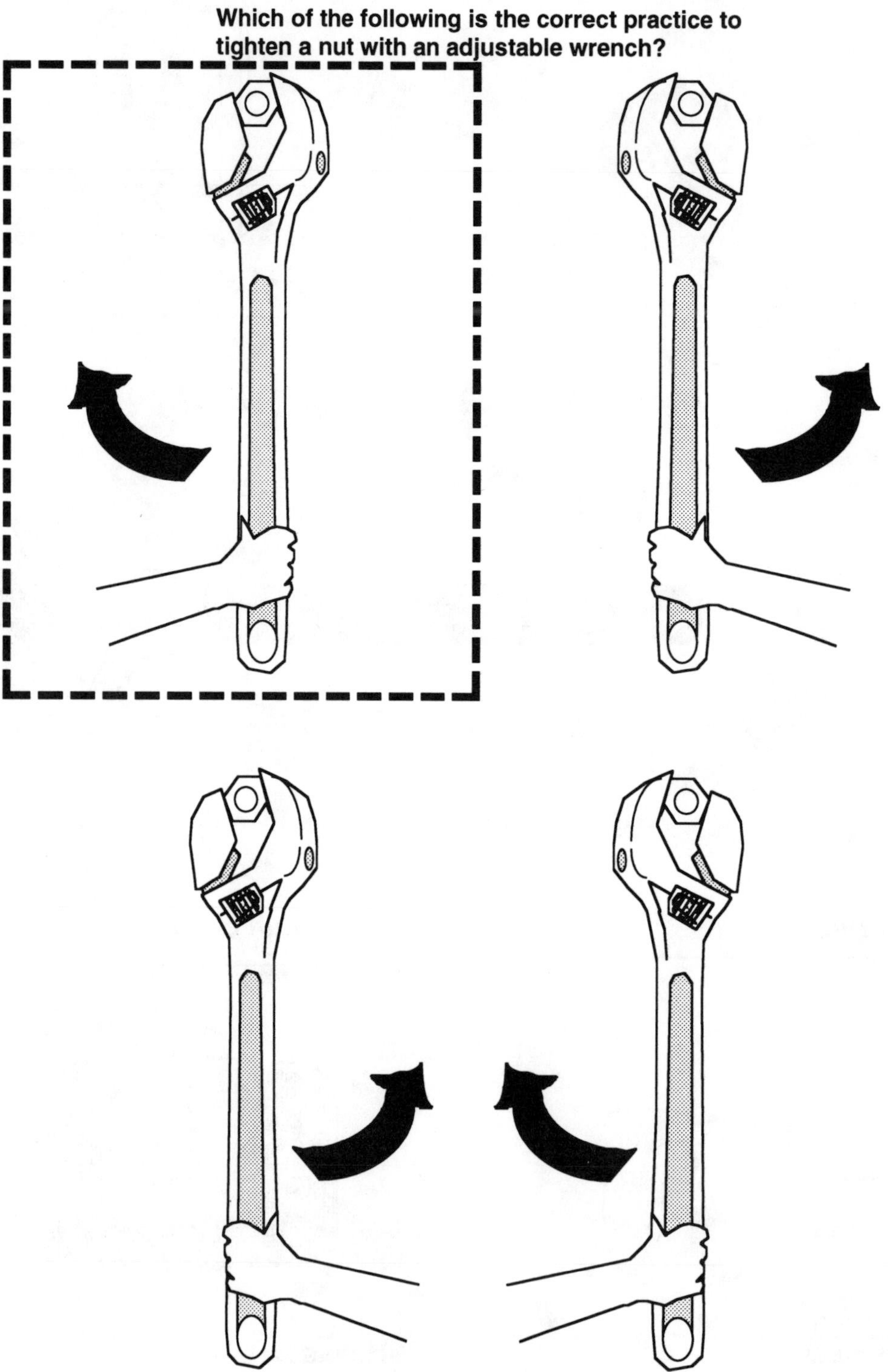

JOURNEYMAN WIREMAN QUIZ #32
PRACTICAL QUESTIONS

•*Circle the correct answer:*

1. The neutral conductor shall not be ____.

(a) stranded
(b) solid
(c) insulated
(d) fused

2. The part of an electrical system that performs a mechanical function rather than an electrical function is called a(n) ____.

(a) receptacle
(b) device
(c) fitting
(d) outlet

3. Solid wire is preferred instead of stranded wire in panel wiring because ____.

(a) costs less than stranded
(b) solid will carry more current
(c) can be "shaped" better
(d) no derating required for solid

4. What is meant by "traveler wires"?

(a) wiring to a split receptacle
(b) two-wires between 3-way switches
(c) wiring to a door bell
(d) out of state electrician

5. When working near acid storage batteries, extreme care should be taken to guard against sparks, essentially to avoid ____.

(a) overheating the electrolyte
(b) an electric shock
(c) a short circuit
(d) an explosion

JOURNEYMAN WIREMAN QUIZ #33
TESTING

Which of the fuses is blown?

- *Circle the line that the fuse is* <u>*BLOWN.*</u> *L1* *or* *L2*

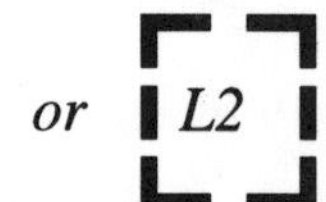

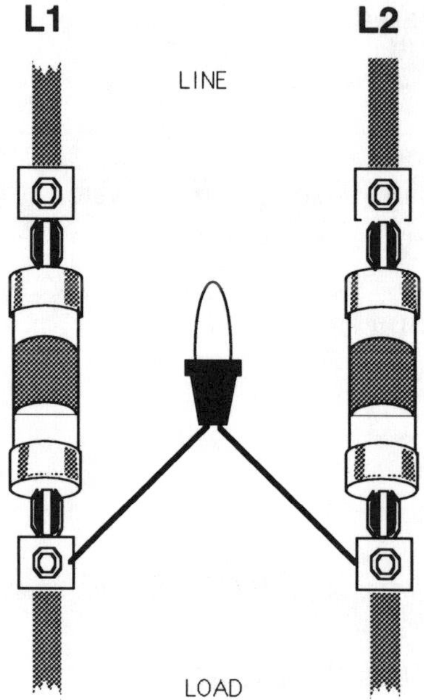

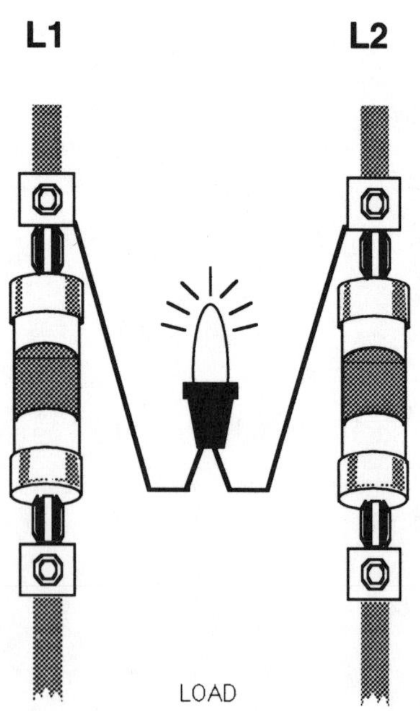

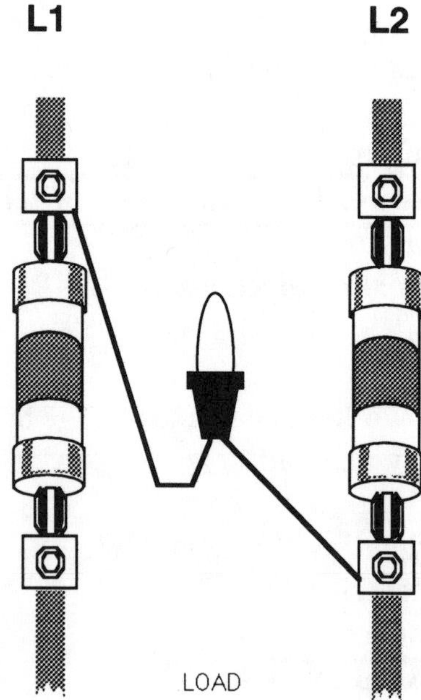

JOURNEYMAN WIREMAN QUIZ #34
INSTALLATION

•Circle the correct installation method.

Which of the following is the correct practice to splice a cord?

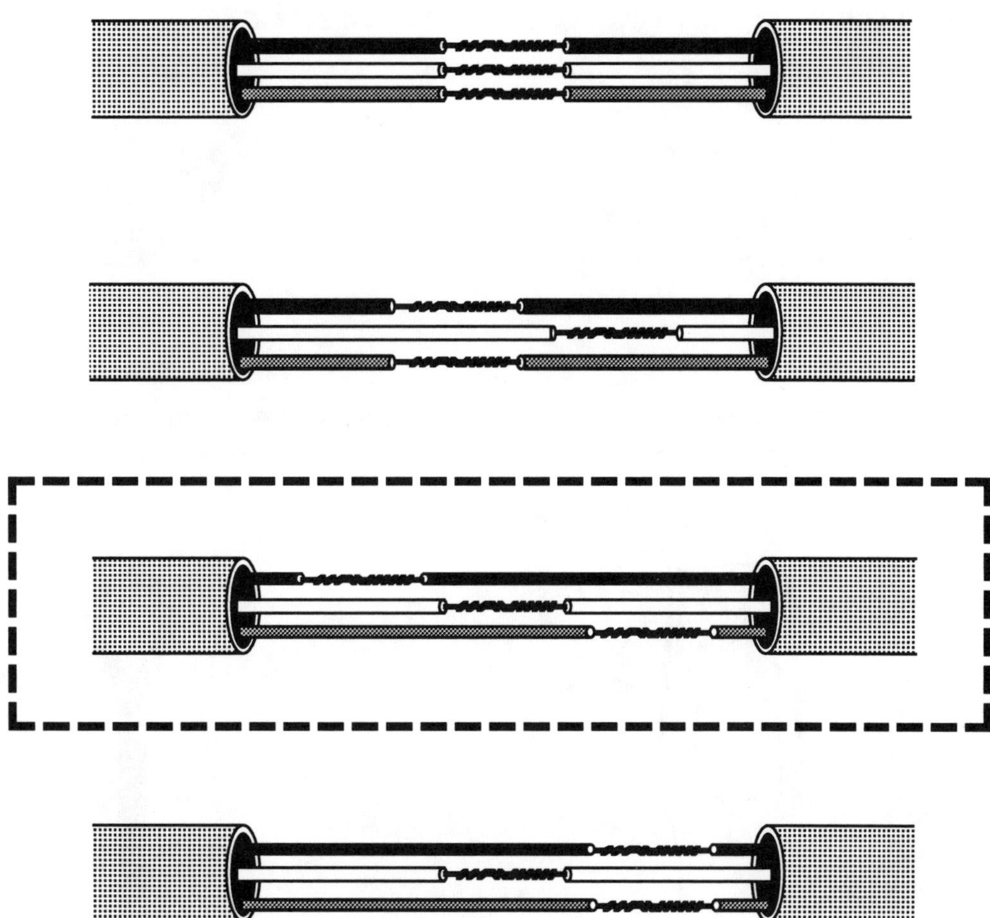

JOURNEYMAN WIREMAN QUIZ #35
TOOL IDENTIFICATION QUESTIONS

• *Fill in the blank with the correct name for the tool.*

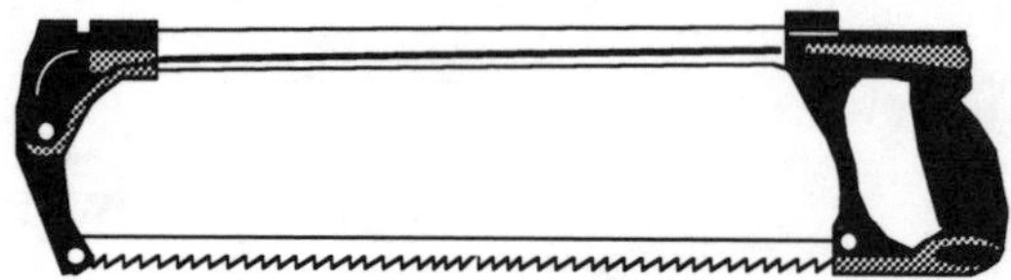

1. **Aviation snips**

2. **Hack saw**

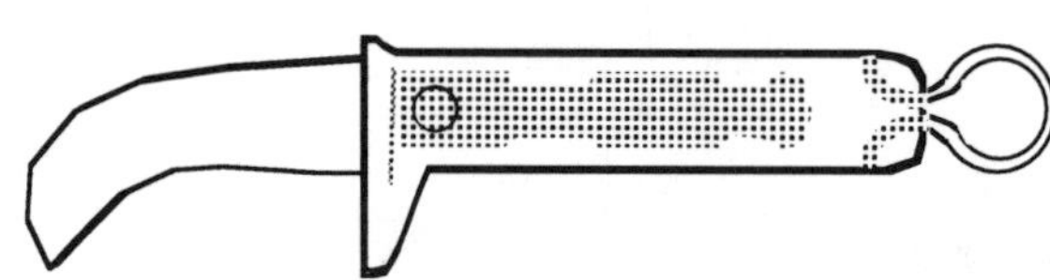

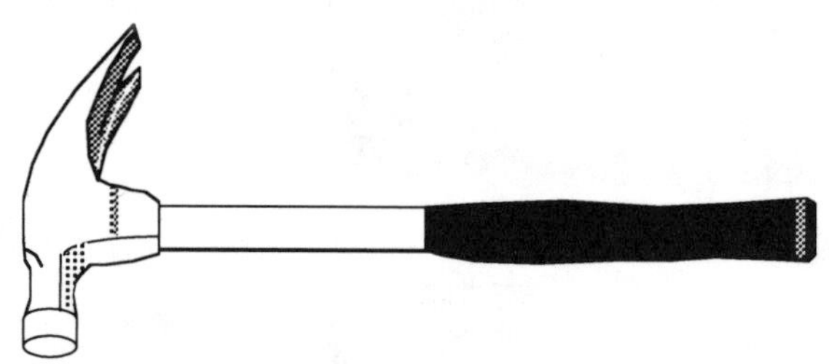

3. **Lineman's skinning knife**

4. **Straight-claw hammer**

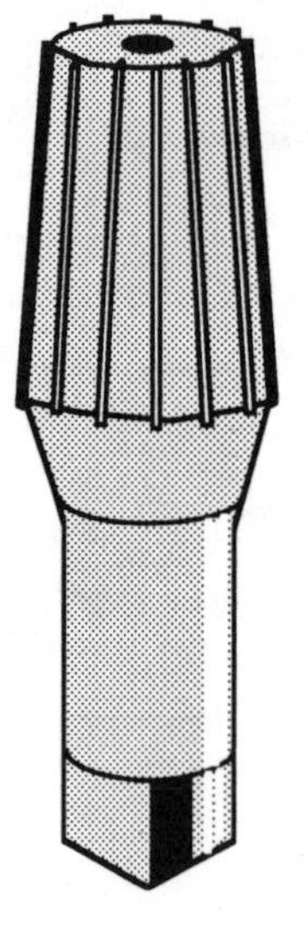

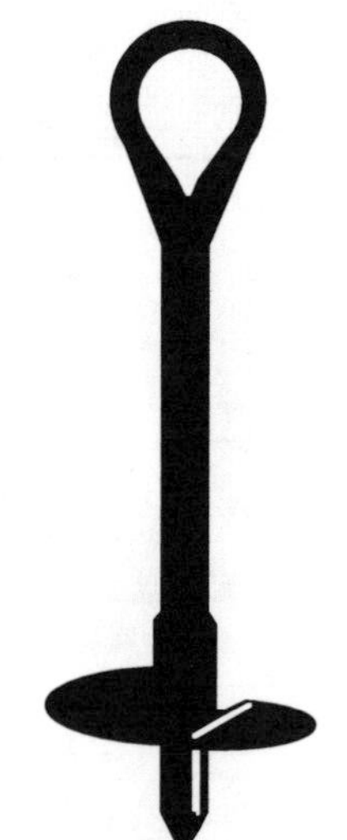

5. **Pipe reamer**

6. **Guy anchor**

JOURNEYMAN WIREMAN QUIZ #36
PRACTICAL QUESTIONS

•*Circle the correct answer:*

1. The reason for grounding the frame of a portable electric hand tool is to ____.

(a) prevent the frame of the tool from becoming alive to ground
(b) prevent overheating of the tool
(c) prevent shorts
(d) reduce voltage drop

2. The purpose of a Western Union splice is ____.

(a) for the use of the utility companies only
(b) for the purpose of strengthening a splice
(c) for the use on the west coast only
(d) none of the above

3. To mark a point on the floor directly beneath a point on the ceiling, it is best to use a ____.

(a) transit rod
(b) plumb bob
(c) square
(d) 12' tape

4. When installing an instrument meter on a panel, to obtain accurate mounting ____.

(a) use the meter and drill thru the holes
(b) drill oversize holes
(c) use a template
(d) drill from back of panel

5. The advantage of cutting a metal rigid conduit with a hacksaw rather than a pipe cutter is ____.

(a) you do not need a vice
(b) less energy required in cutting
(c) less reaming is required
(d) threading oil is not required

JOURNEYMAN WIREMAN QUIZ #37
CONTROL SYMBOLS

•*Fill in the blank with the correct letter from choices below for the symbol*

1. **J**
2. **F**
3. **E**
4. **K**
5. **Q**
6. **I**

7. **H**
8. **P**
9. **D**
10. **S**
11. **B**
12. **T**

13. **O**
14. **C**
15. **G**
16. **L**
17. **M**
18. **N**

19. **A**
20. **R**

•*Choose a letter ( ) and fill in the blank above:*

(A) CB with thermal O.L.
(B) normally closed contact
(C) liquid level switch N.C.
(D) temperature actuated switch N.O.
(E) foot switch N.C.
(F) start button N.O.
(G) Timed contact N.O.T.C.
(H) stop button N.C.
(I) disconnect
(J) limit switch N.O.
(K) SPDT double break
(L) normally open contact
(M) temperature actuated switch N.C.
(N) thermal O.L.
(O) selector switch-two position
(P) foot switch N.O.
(Q) limit switch N.C.
(R) autotransformer winding
(S) liquid level switch N.O.
(T) mushroom head push button switch

JOURNEYMAN WIREMAN QUIZ #38
TOOL IDENTIFICATION QUESTIONS

• *Fill in the blank with the correct name for the tool.*

1. CRIMPING TOOL

2. HEAT GUN

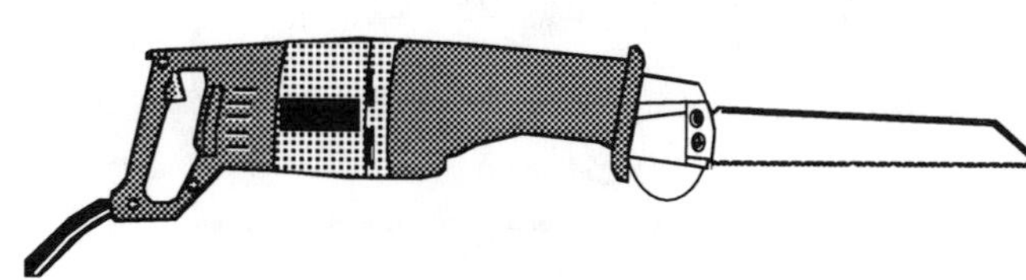

3. SAWZALL

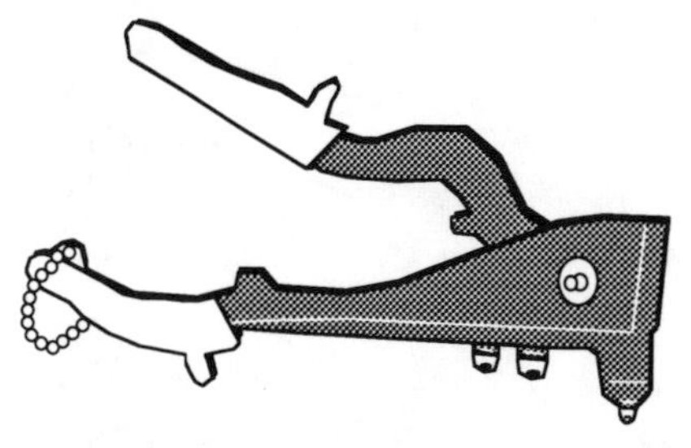

4. RIVETING TOOL

5. ROTOMETER

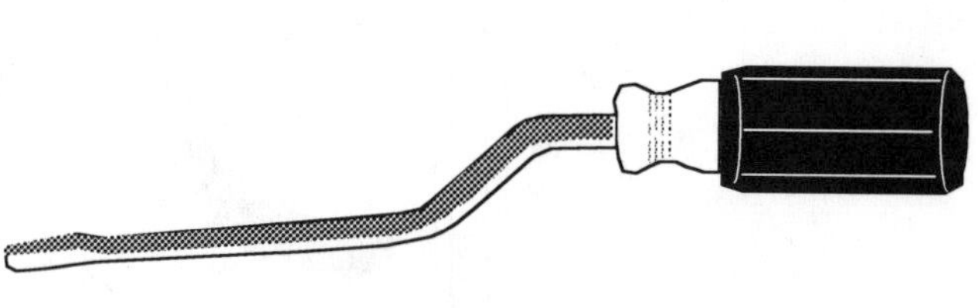

6. ROTARY SCREWDRIVER

JOURNEYMAN WIREMAN QUIZ #39
INSTALLATION

•Circle the correct installation method.

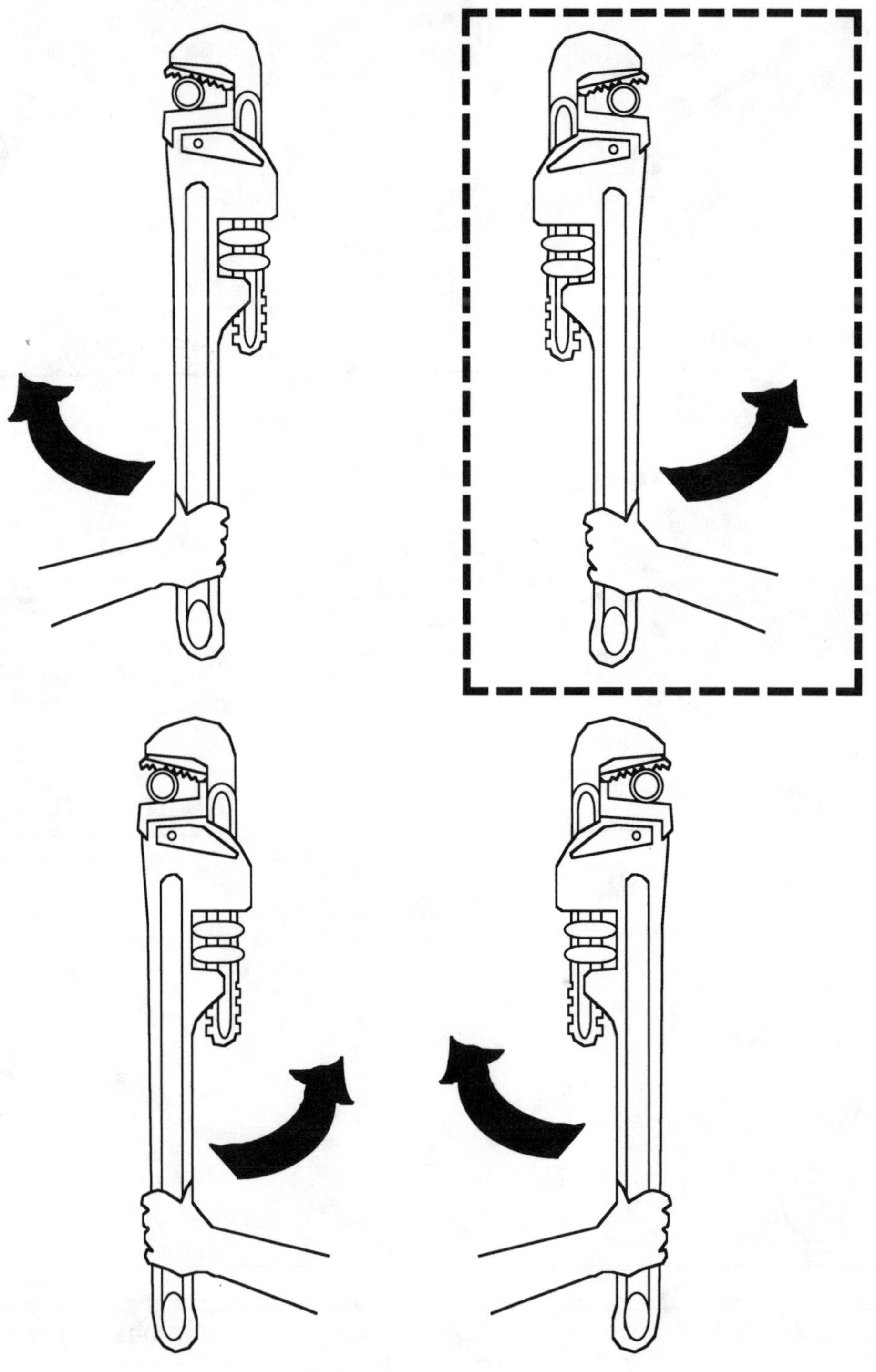

JOURNEYMAN WIREMAN QUIZ #40
TOOL IDENTIFICATION QUESTIONS

• *Fill in the blank with the correct name for the tool.*

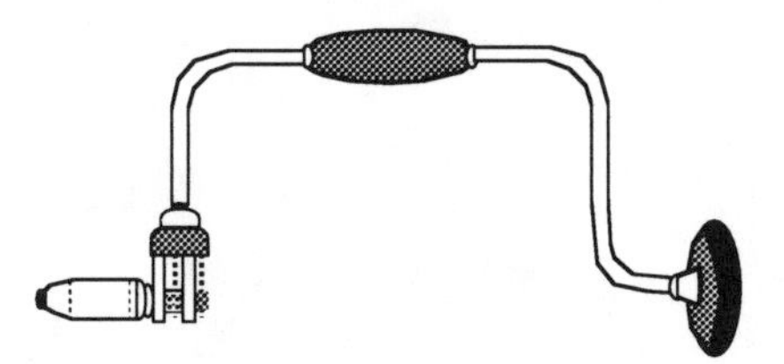

1. BIT BRACE

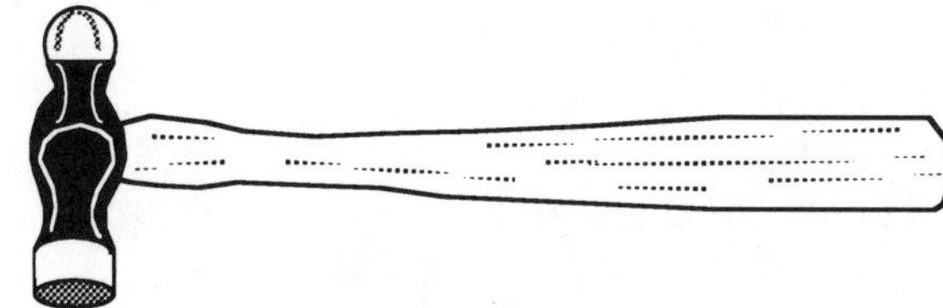

2. BALL PEAN HAMMER

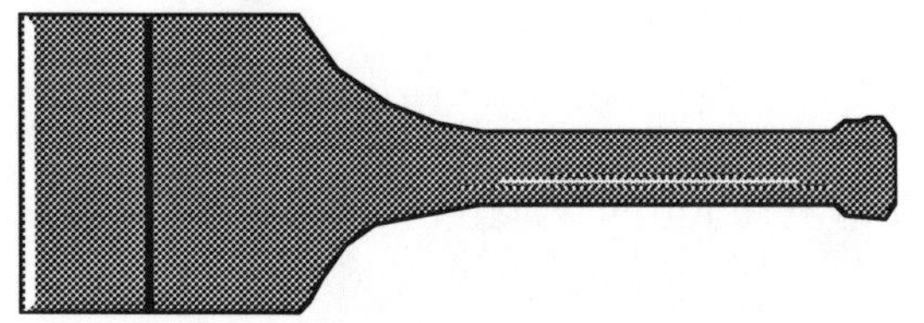

3. ELECTRICIAN'S CHISEL

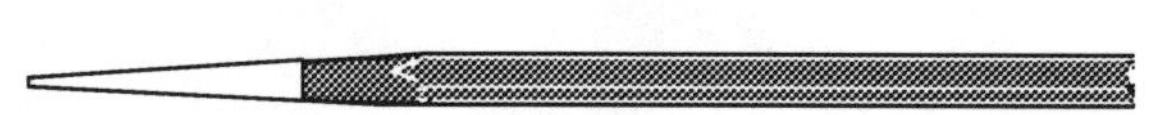

4. DRIFT PUNCH

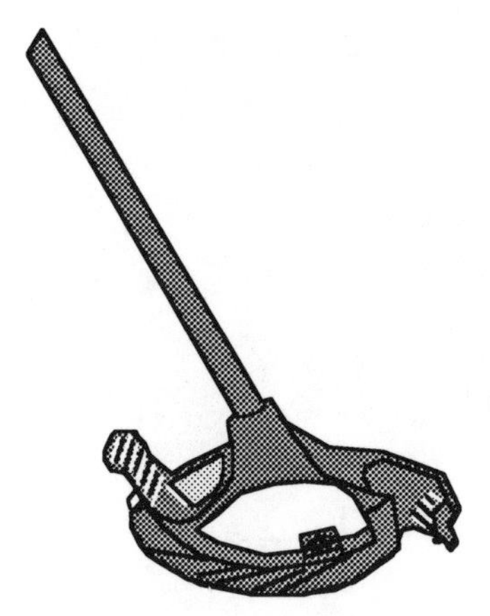

5. CONDUIT BENDER

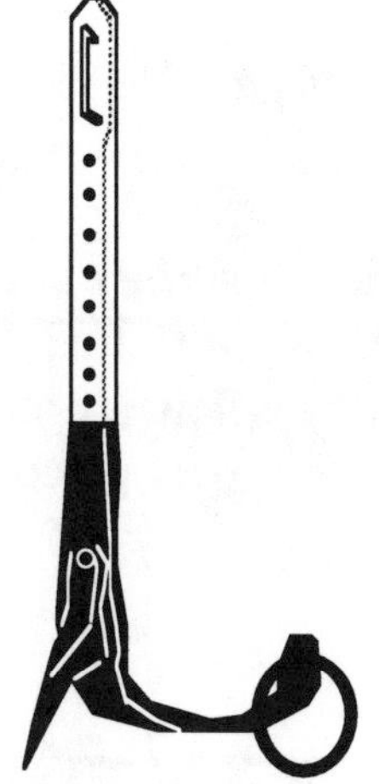

6. POLE CLIMBER

JOURNEYMAN WIREMAN QUIZ #41
PRACTICAL QUESTIONS

•*Circle the correct answer:*

1. Multiple start buttons in a motor control circuit are connected in ____.

(a) series
(b) parallel
(c) series-parallel
(d) none of the above

2. A function of a relay is to ____.

(a) turn on another circuit
(b) produce thermal electricity
(c) limit the flow of electrons
(d) create a resistance in the field winding

3. To control a ceiling light from five different locations it requires which of the following?

(a) four 3-way switches and one 4-way switch
(b) three 4-way switches and two 3-way switches
(c) three 3-way and two 4-way switches
(d) four 4-way switches and one 3-way switch

4. The identified grounded conductor of a lighting circuit is always connected to the screw of a light socket to ____.

(a) reduce the possibility of accidental shock
(b) ground the light fixture
(c) improve the efficiency of the lamp
(d) provide the easiest place to connect the wire

5. If the end of a cartridge fuse becomes warmer than normal, you should ____.

(a) tighten the fuse clips
(b) lower the voltage on the circuit
(c) notify the ultility company
(d) change the fuse

JOURNEYMAN WIREMAN QUIZ #42
WIRING METHODS

• *Fill in the blank with the correct letter for the wiring method or equipment:*

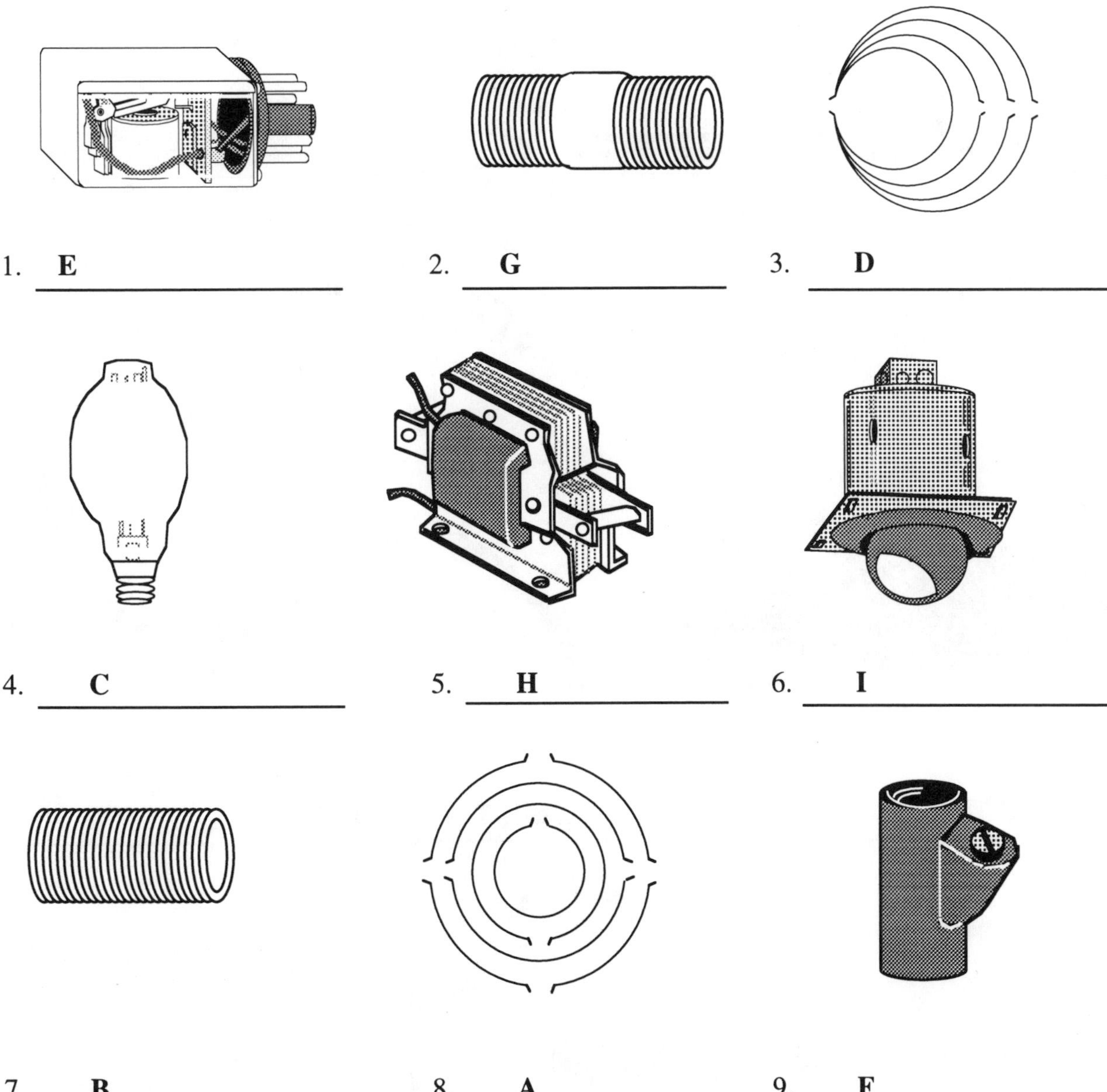

1. **E**
2. **G**
3. **D**
4. **C**
5. **H**
6. **I**
7. **B**
8. **A**
9. **F**

• *Choose a letter ( ) and fill in the blank above:*

(A) concentric K.O.
(B) close nipple
(C) mercury lamp
(D) eccentric K.O.
(E) relay
(F) seal off fitting
(G) short nipple
(H) solenoid
(I) bull eye's light

JOURNEYMAN WIREMAN QUIZ #43
TOOL IDENTIFICATION QUESTIONS

• *Fill in the blank with the correct name for the tool.*

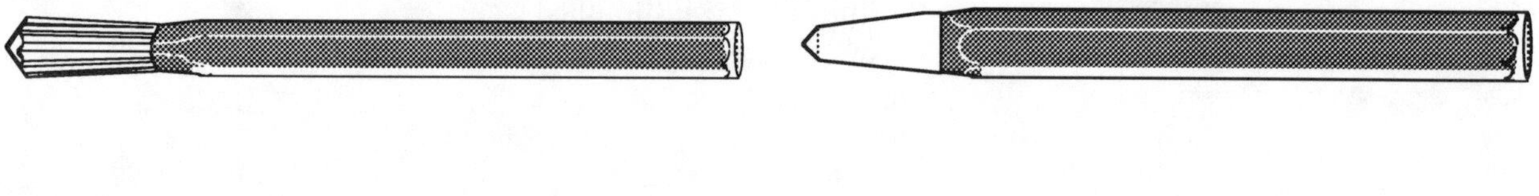

1. STAR DRILL

2. CENTER PUNCH

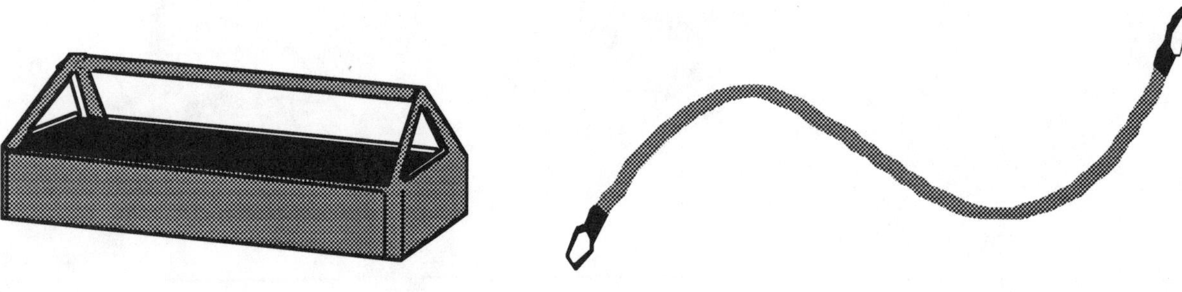

3. TOTE TRAY

4. FISH-TAPE LEADER

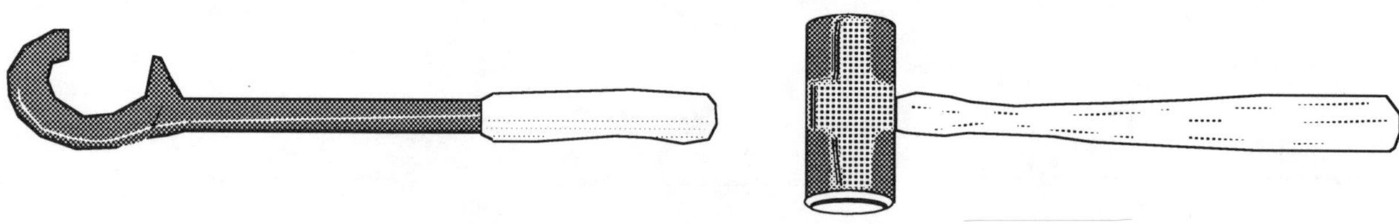

5. CABLE BENDER

6. SLEDGE HAMMER

JOURNEYMAN WIREMAN QUIZ #44
INSTALLATION

•Circle the correct installation method.

Which of the following is the correct practice when cutting thin wall conduit?

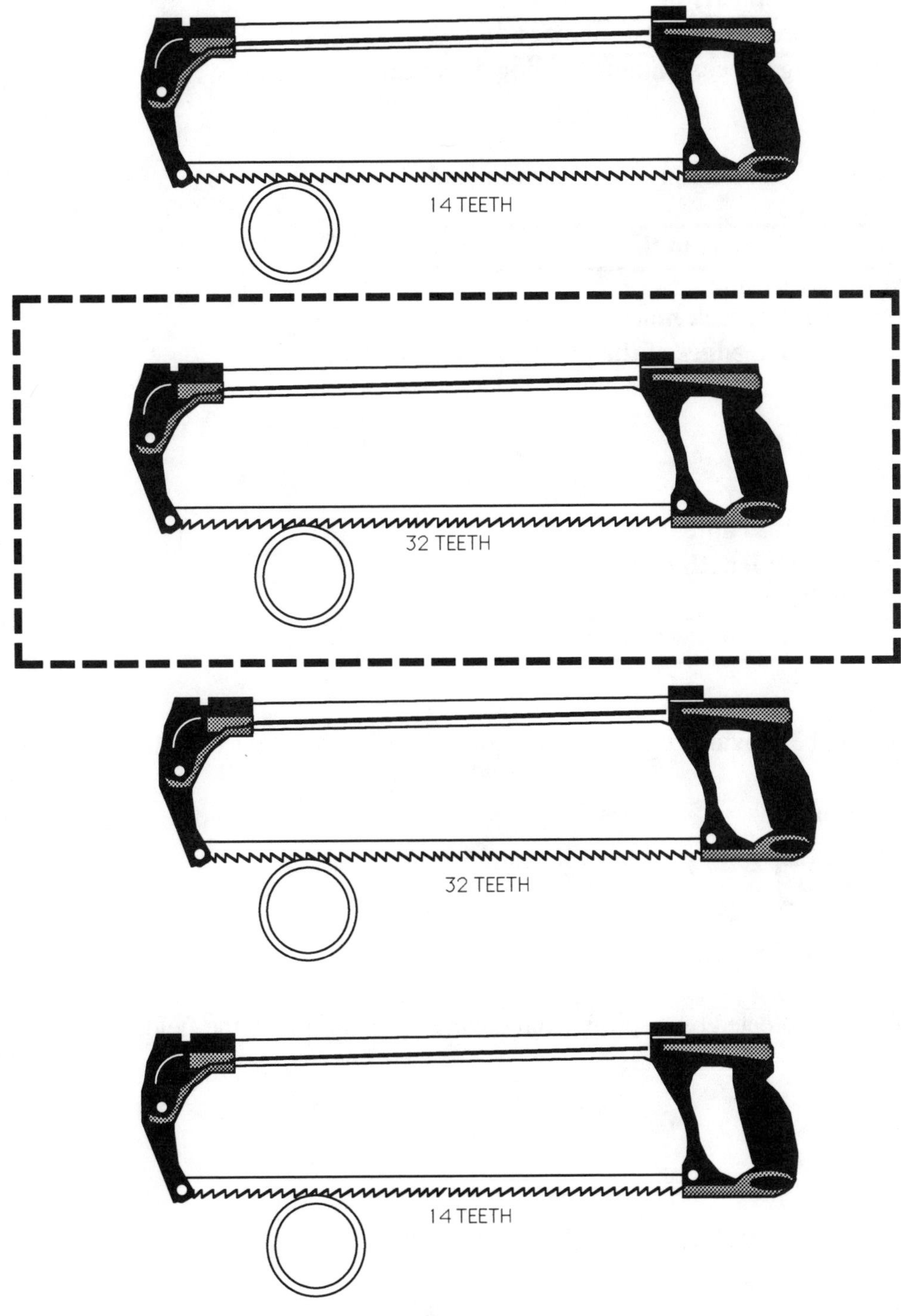

JOURNEYMAN WIREMAN QUIZ #45
PRACTICAL QUESTIONS

•Circle the correct answer:

1. The reason for installing electrical conductors in a conduit is _____.

(a) to provide a ground
(b) to increase the ampacity of the conductors
(c) to protect the conductors from damage
(d) to avoid derating for continuous loading of conductors

2. A ladder which is painted is a safety hazard mainly because the paint _____.

(a) may conceal weak spots in the rails or rungs
(b) is slippery after drying
(c) causes the wood to crack more quickly
(d) peels and the sharp edges of the paint may cut the hands

3. A conduit body is ____.

(a) a cast fitting such as an FD or FS box
(b) a standard 10 foot length of conduit
(c) a sealtight enclosure
(d) an "LB" or "T", or similar fitting

4. Raceways shall be provided with ____ to compensate for thermal expansion and contraction.

(a) accordion joints
(b) thermal fittings
(c) expansion joints
(d) contro-spansion

5. A type of cable protected by a spiral metal cover is called ____ in the field.

(a) BX
(b) greenfield
(c) sealtight
(d) Romex

JOURNEYMAN WIREMAN QUIZ #46
WIRING METHODS

• *Fill in the blank with the correct letter for the wiring method or equipment:*

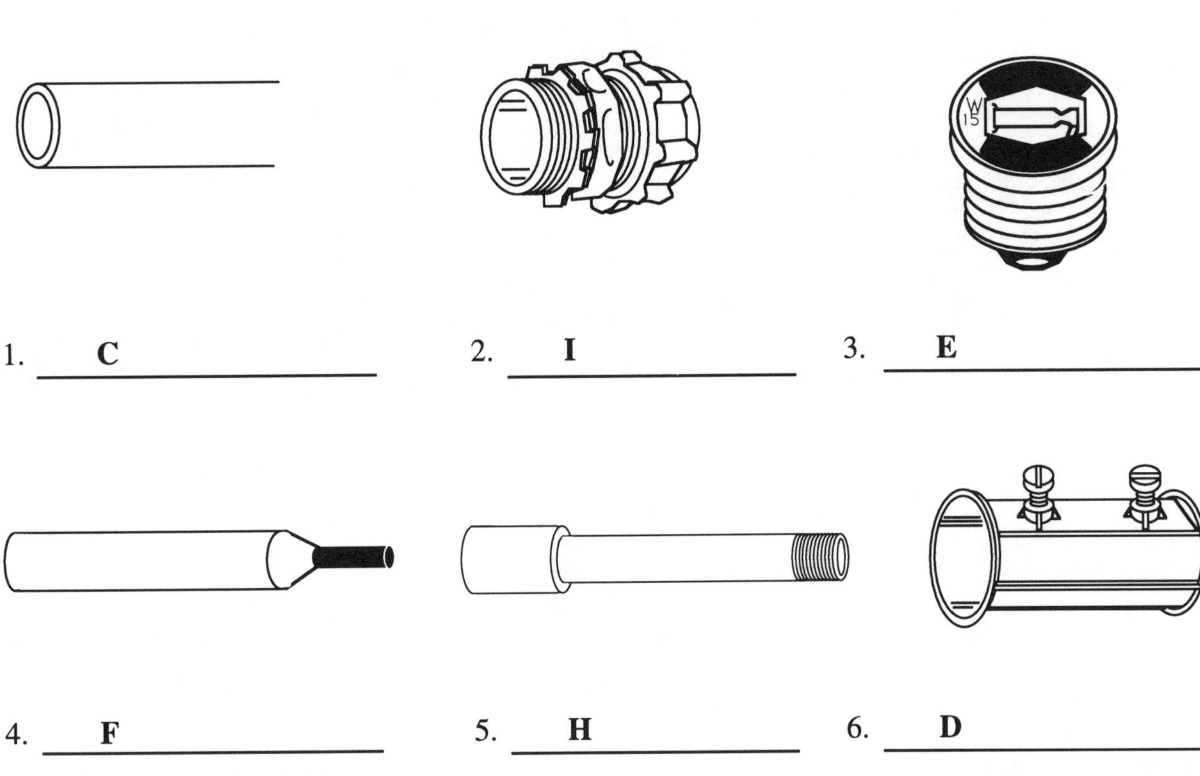

1. **C** __________ 2. **I** __________ 3. **E** __________

4. **F** __________ 5. **H** __________ 6. **D** __________

7. **G** __________ 8. **B** __________ 9. **A** __________

• *Choose a letter ( ) and fill in the blank above:*

(A) knife-blade fuse
(B) wrong way to skin a wire
(C) thin wall tubing
(D) coupling
(E) plug fuse
(F) right way to skin a wire
(G) ferrule-contact fuse
(H) rigid conduit
(I) box connector

JOURNEYMAN WIREMAN QUIZ #47
TOOL IDENTIFICATION QUESTIONS

• *Fill in the blank with the correct name for the tool.*

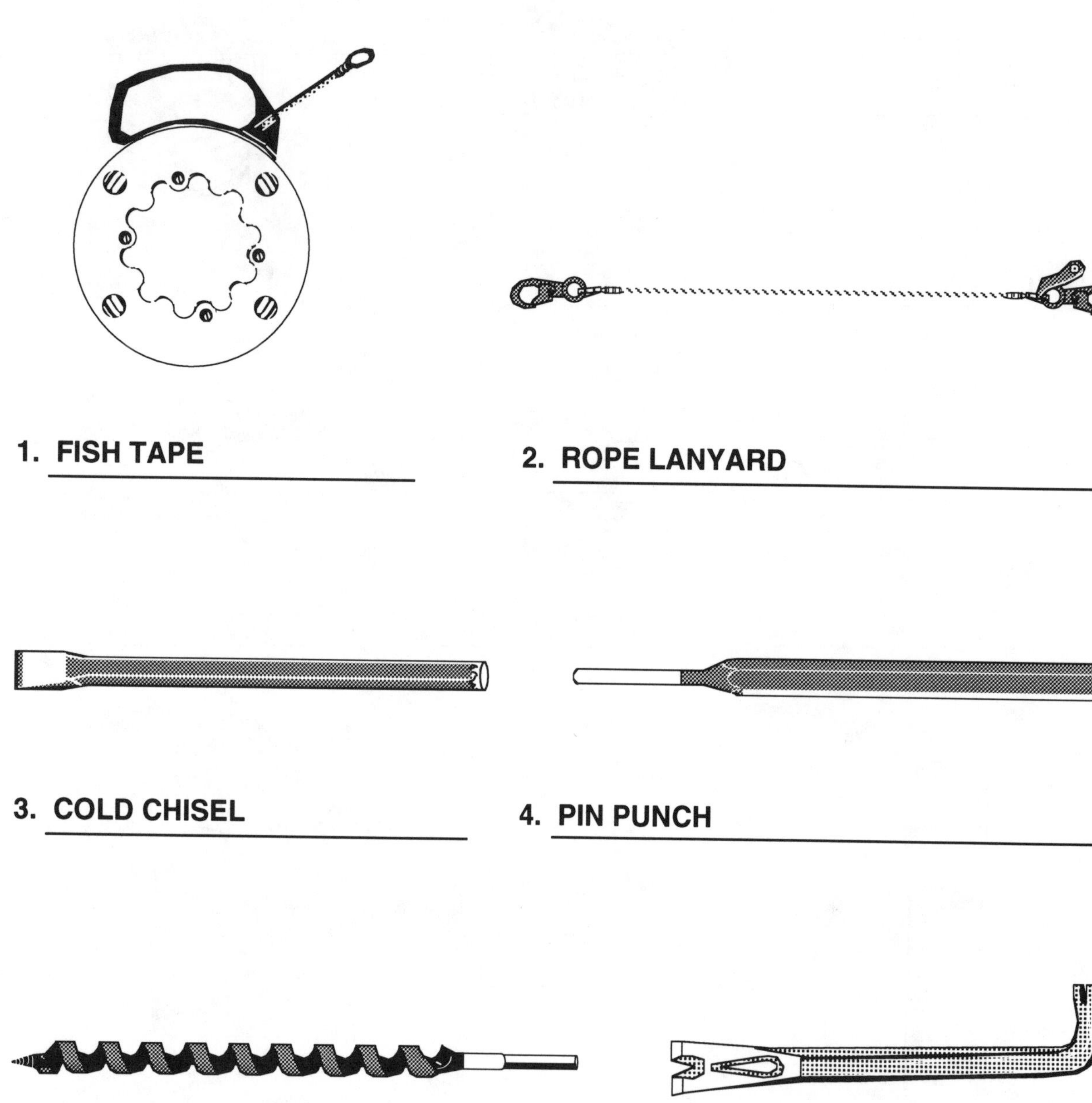

1. FISH TAPE

2. ROPE LANYARD

3. COLD CHISEL

4. PIN PUNCH

5. AUGER DRILL BIT

6. RIPPING BAR

JOURNEYMAN WIREMAN QUIZ #48
INSTALLATION

•Circle the correct installation method.

Which of the following is the correct practice to crush the insulation prior to skinning the wire?

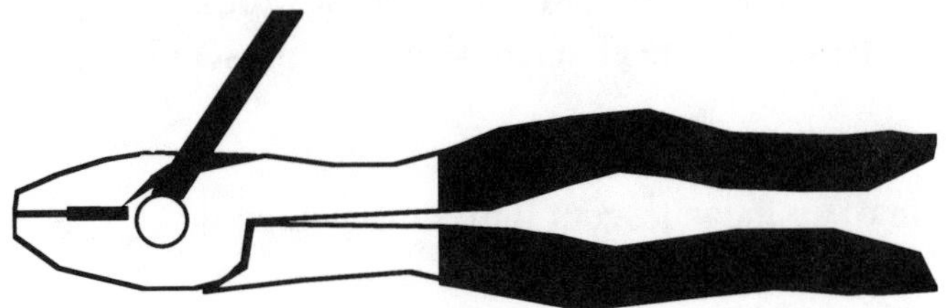

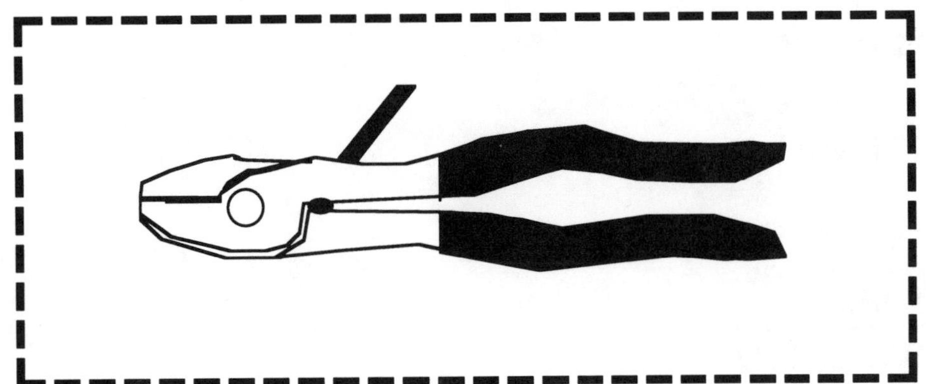

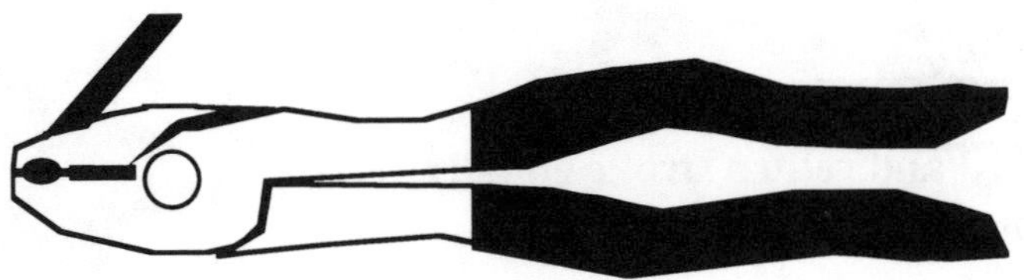

JOURNEYMAN WIREMAN QUIZ #49
PRACTICAL QUESTIONS

•*Circle the correct answer:*

1. With respect to fluorescent lamps it is correct to state ____.

(a) the filaments seldom burn out
(b) the starters and tubes must be replaced at the same time
(c) they are easier to install than incandescent light bulbs
(d) their efficiency is less than the efficiency of incandescent light bulbs

2. To increase the life of an incandescent light bulb you could ____.

(a) use at a higher than rated voltage
(b) use at a lower than rated voltage
(c) turn off when not in use
(d) use at a higher wattage

3. Which of the following hacksaw blades should be used for the best results in cutting EMT?

(a) 12 teeth per inch
(b) 18 teeth per inch
(c) 24 teeth per inch
(d) 32 teeth per inch

4. The letters DPDT are used to identify a type of ____.

(a) insulation
(b) fuse
(c) motor
(d) switch

5. When cutting a metal conduit with a hacksaw, the pressure applied to the hacksaw should be on ____.

(a) the return stroke
(b) the forward stroke only
(c) both the forward and return stroke equally
(d) none of the above

JOURNEYMAN WIREMAN QUIZ #50
TOOL IDENTIFICATION QUESTIONS

• *Fill in the blank with the correct name for the tool.*

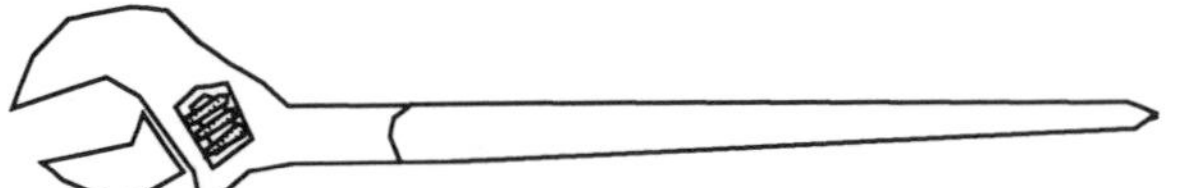

1. CONSTRUCTION WRENCH

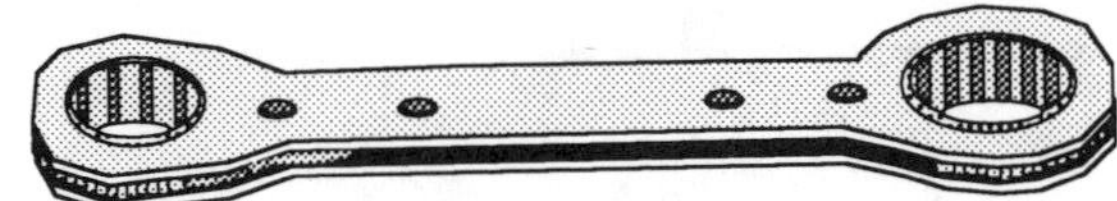

2. RATCHETING BOX WRENCH

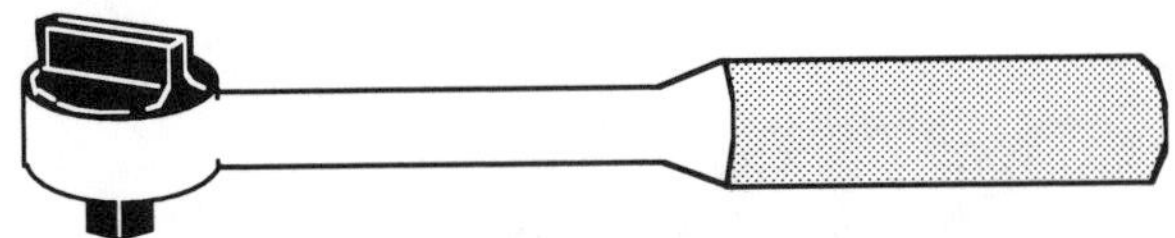

3. SOCKET WRENCH

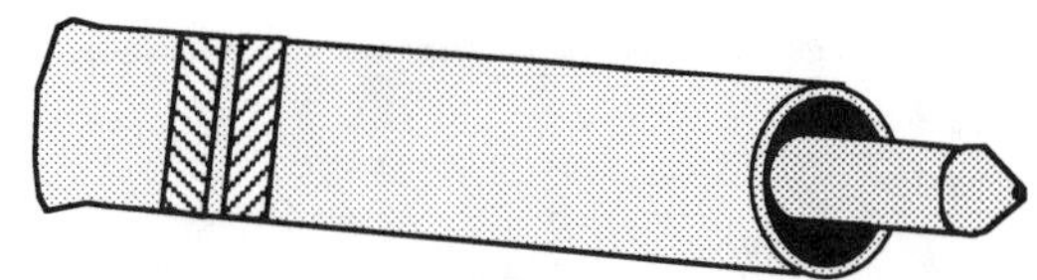

4. LEAD ANCHOR SET

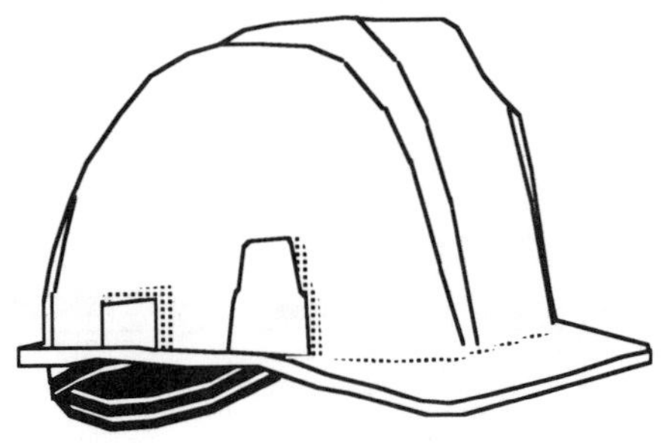

5. HARD HAT

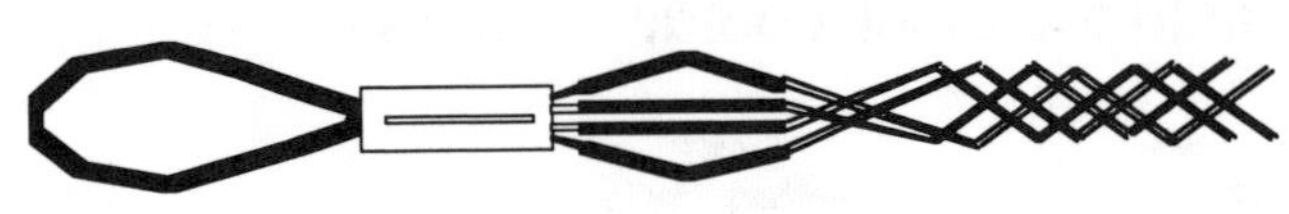

6. WIRE MESH CABLE GRIP

JOURNEYMAN WIREMAN QUIZ #51
INSTALLATION

•Circle the correct installation method.

Which of the following is the correct practice when installing two wires to a stud using flat washers and hex nuts?

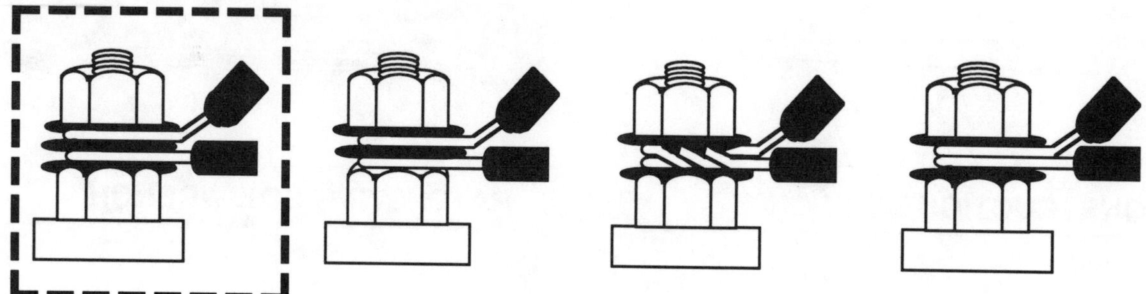

Which of the following is the correct practice when installing a wire around a binding post?

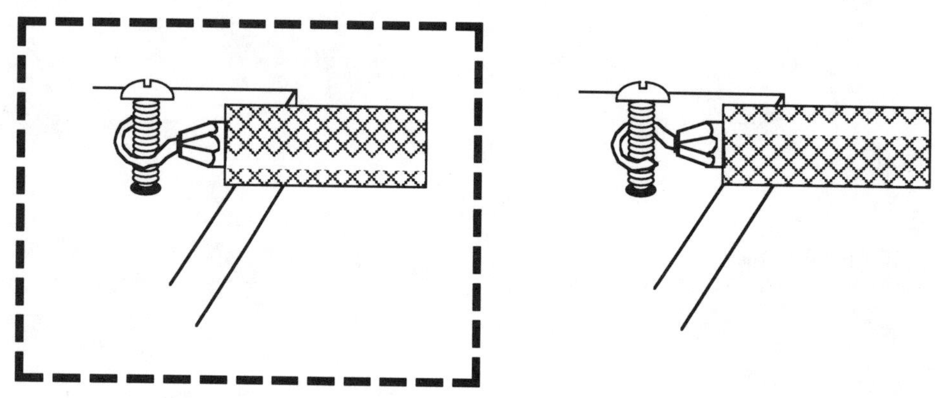

Fill in the blanks below naming the splices shown

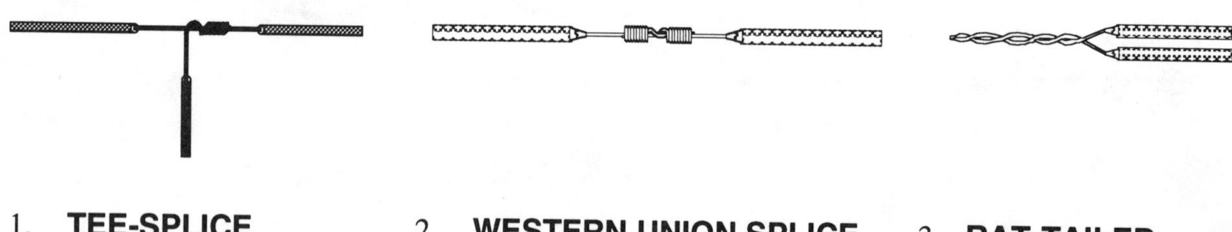

1. **TEE-SPLICE**
2. **WESTERN UNION SPLICE**
3. **RAT-TAILED**

JOURNEYMAN WIREMAN QUIZ #52
BLUEPRINT SYMBOLS

•Fill in the blank with the correct letter from choices below for the symbol

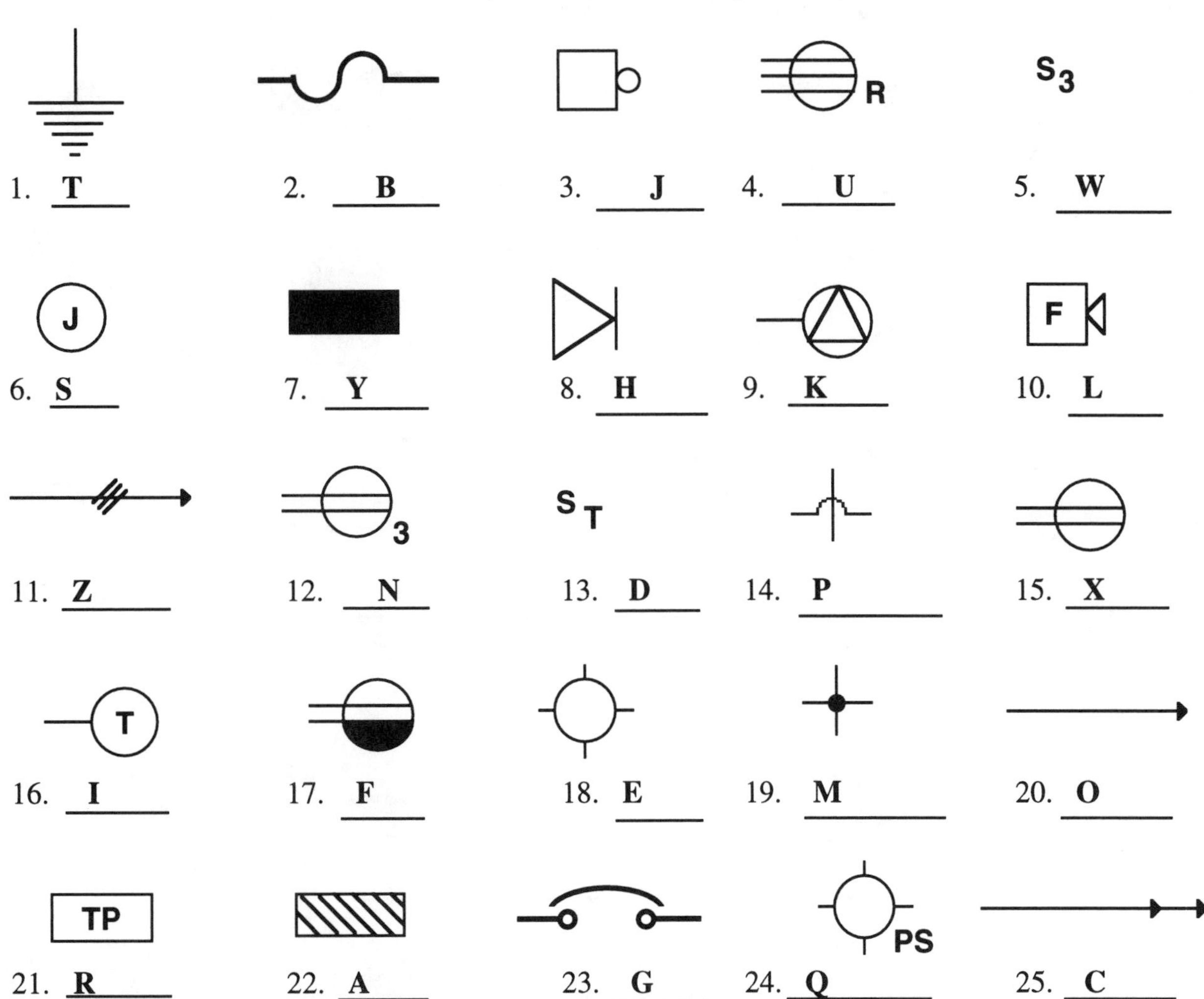

• *Choose a letter ( ) and fill in the blank above:*

(A) power panel
(B) fusible element
(C) two branch circuit home runs to panel
(D) time switch
(E) ceiling outlet
(F) duplex outlet, split circuit
(G) circuit breaker
(H) telephone
(I) thermostat
(J) fire alarm bell
(K) single special-purpose receptacle
(L) fire alarm horn
(M) wiring connected
(N) triplex receptacle outlet
(O) single branch circuit home run to panel
(P) wiring crossed not connected
(Q) lampholder with pull switch
(R) transformer pad
(S) junction box
(T) ground
(U) range receptacle
(W) switch 3-way
(X) duplex receptacle
(Y) branch circuit lighting panel
(Z) single branch circuit home run to panel (3-wire)

JOURNEYMAN WIREMAN #53
TOOL IDENTIFICATION QUIZ

•*Fill in the blank with the correct letter from choices below for the tool shown:*

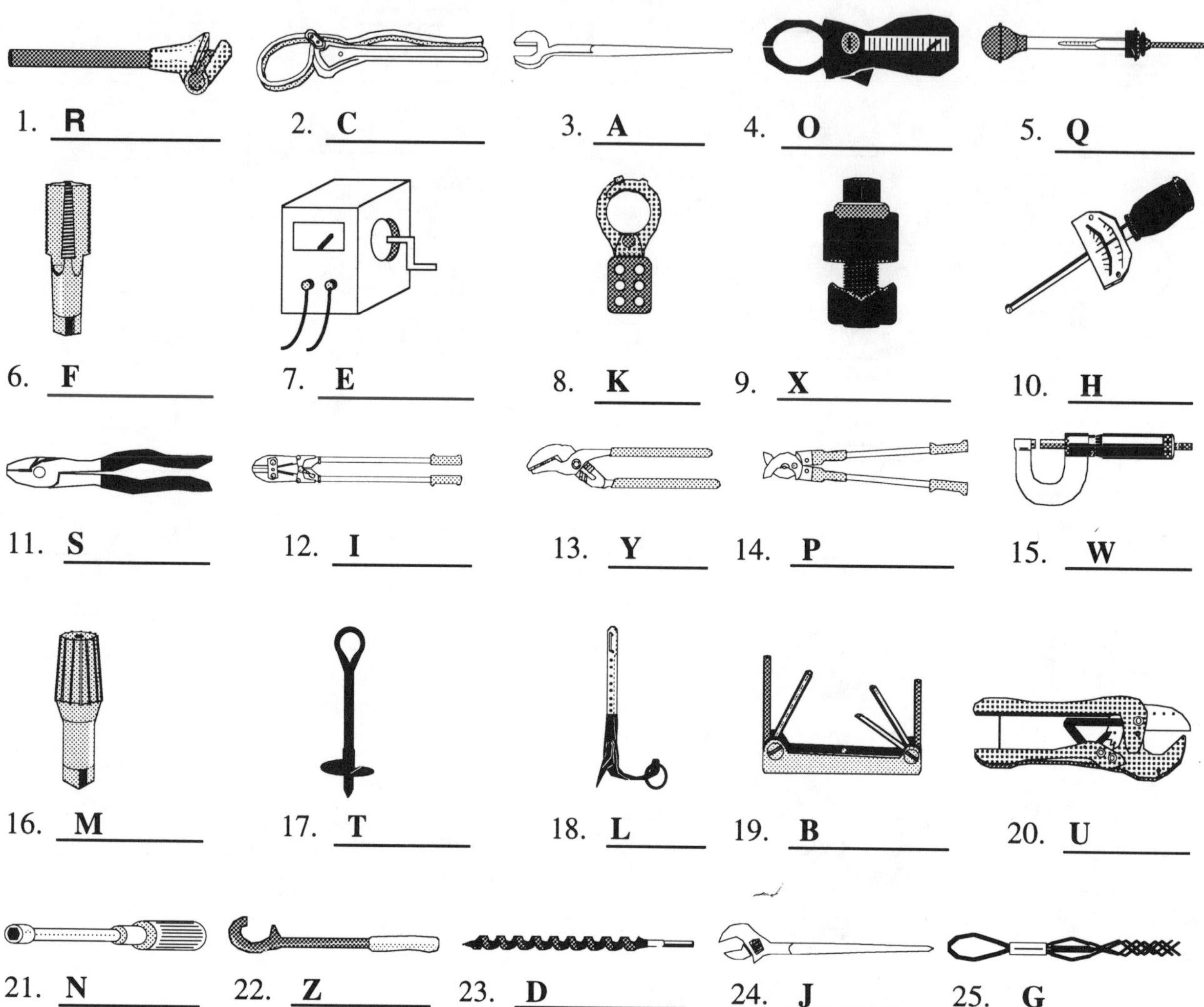

•*Choose a letter ( ) and fill in the blank above:*

(A) erection wrench
(B) hex key set
(C) strap wrench
(D) auger drill bit
(E) megger
(F) pipe thread tap
(G) wire mesh cable grip
(H) torque wrench
(I) bolt cutters
(J) construction wrench
(K) lock out
(L) pole climbers
(M) pipe reamer
(N) nut driver
(O) clamp on meter
(P) cable cutters
(Q) hydrometer
(R) hickey
(S) side cutting pliers
(T) guy anchor
(U) PVC cutters
(W) micrometer
(X) knock out
(Y) pump pliers
(Z) cable bender

JOURNEYMAN WIREMAN QUIZ #54
WIRING METHODS QUIZ

• *Fill in the blank with the correct letter for the wiring method or equipment:*

1. **J** 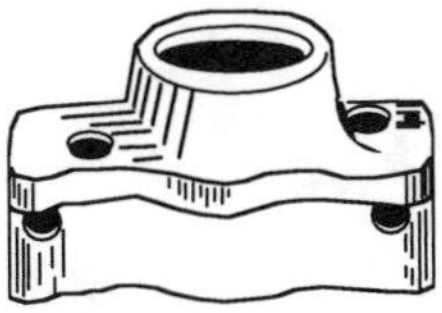
2. **O** 
3. **M** 
4. **B** 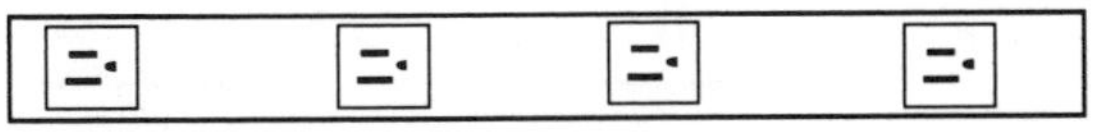
5. **N** 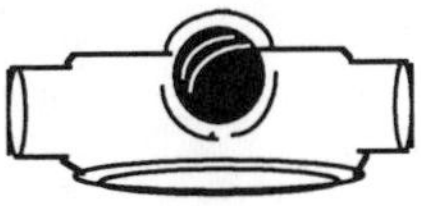
6. **E** 
7. **G** 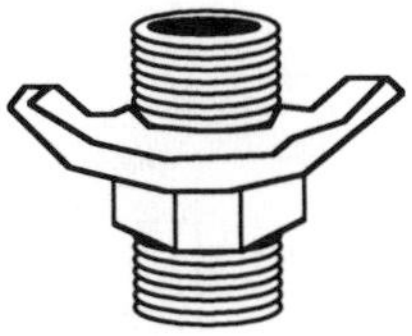
8. **F** 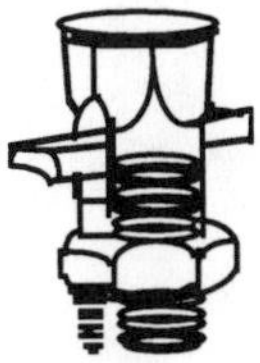
9. **H** 
10. **L** 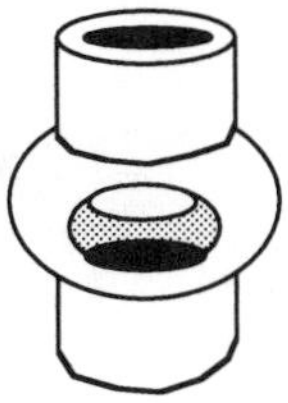
11. **A**

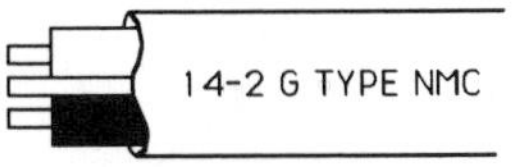

12. **I**
13. **C** 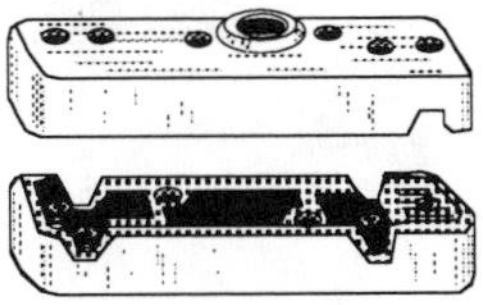
14. **D** 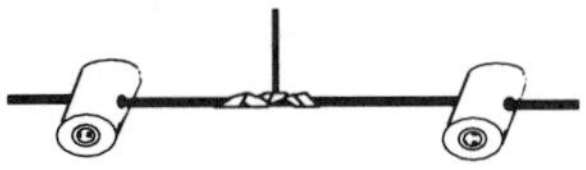
15. **K** 

• *Choose a letter ( ) and fill in the blank above:*

(A) hickey
(B) double-pole switch
(C) rosette
(D) open-wiring
(E) conduit tee
(F) fixture stud
(G) two-screw connector
(H) split-bolt connector
(I) Romex
(J) armored cable
(K) four-way switch
(L) wire nut
(M) cable clamp
(N) plugmold
(O) open-wiring receptacle

JOURNEYMAN WIREMAN QUIZ #55
BLUEPRINT SYMBOLS

•Fill in the blank with the correct letter from choices below for the symbol

S_K | WP | F | S

1. **W** 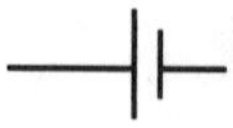
2. **K** 
3. **N** 
4. **G** 
5. **O**

S_P | D | M

6. **S** 
7. **A** 
8. **E** 
9. **B** 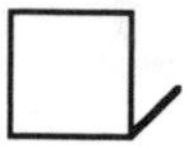
10. **J**

S_F | H

11. **H**
12. **F**
13. **C** 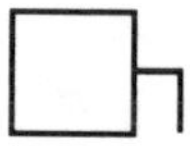
14. **L** 
15. **X**

R | B | F

16. **D** 
17. **M** 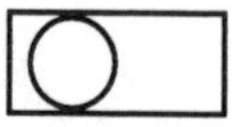
18. **T** 
19. **P** 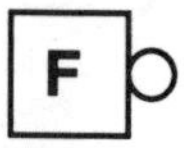
20. **I**

GR

21. **Y**
22. **Q** 
23. **U** 
24. **R**

• Choose a letter ( ) and fill in the blank above:

(A) controller
(B) motor
(C) bare-lamp fluorescent strip
(D) motor starter
(E) electric door opener
(F) switch fused
(G) fan outlet
(H) push button
(I) fire alarm bell
(J) buzzer
(K) one cell
(L) safety switch
(M) fixture recessed
(N) weatherproof outlet
(O) ceiling pull switch
(P) blanked outlet
(Q) grounded duplex receptacle
(R) wiring or conduit turned down
(S) switch with pilot
(T) fluorescent fixture
(U) heating panel
(W) switch key operated
(X) humidistat
(Y) wiring or conduit turned up

JOURNEYMAN QUIZ #56
TOOL IDENTIFICATION QUIZ

•Fill in the blank with the correct letter from choices below for the tool shown:

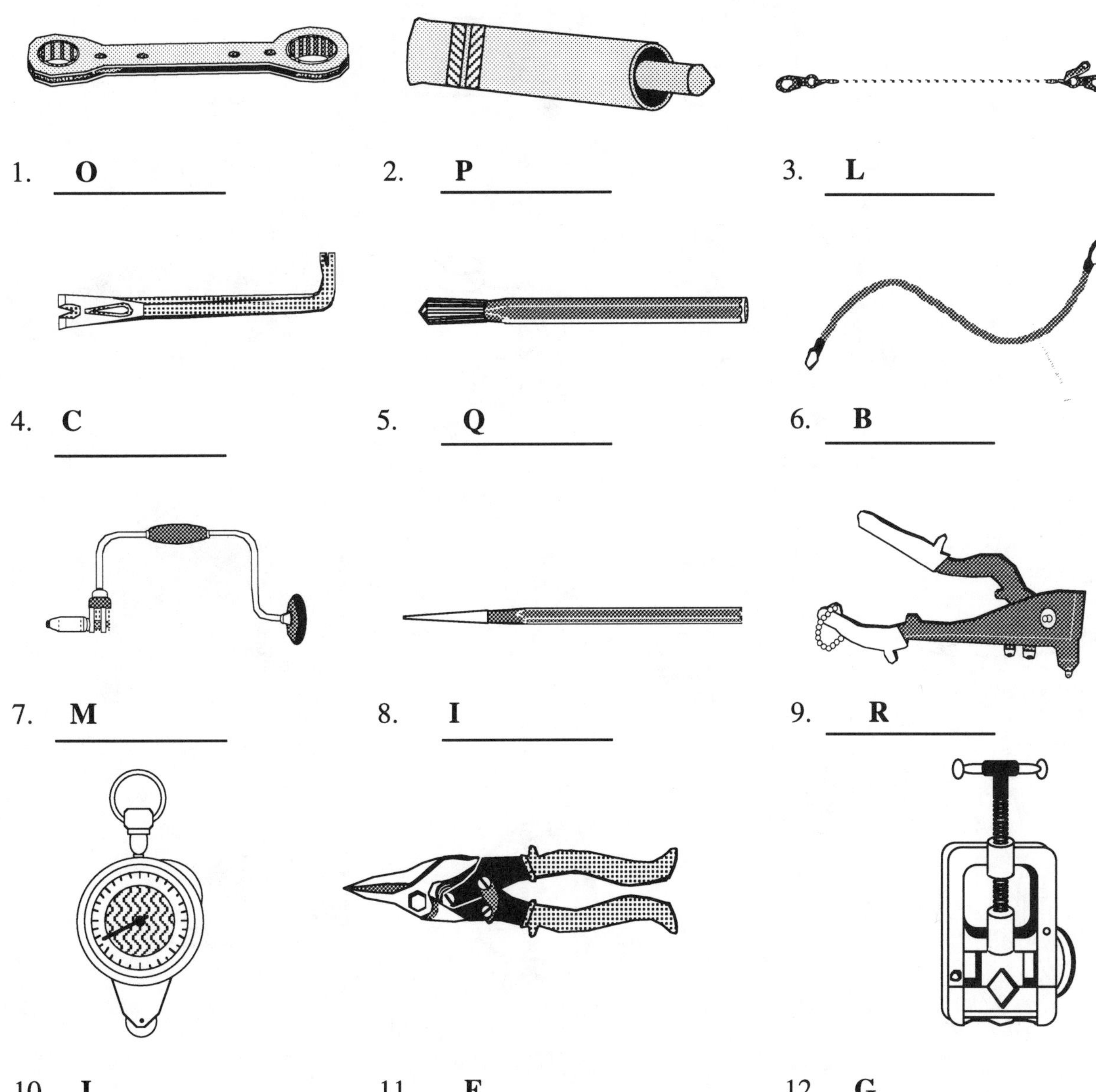

• Choose a letter ( ) and fill in the blank above:

(A) box end wrench
(B) fish-tape leader
(C) ripping bar
(D) hydrometer
(E) aviation snips
(F) lineman pliers
(G) pipe vise
(H) center punch
(I) drift punch
(J) rotometer
(K) pipe clamp
(L) rope lanyard
(M) bit brace
(N) auger bit
(O) ratcheting box wrench
(P) lead anchor set
(Q) star drill
(R) riveting tool

JOURNEYMAN WIREMAN QUIZ #57
WIRING METHODS QUIZ

• *Fill in the blank with the correct letter for the wiring method or equipment:*

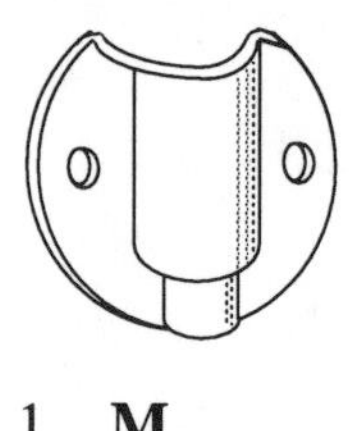

1. **M**

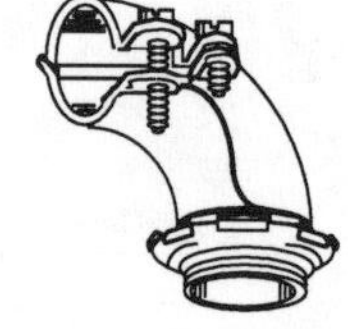

2. **K**

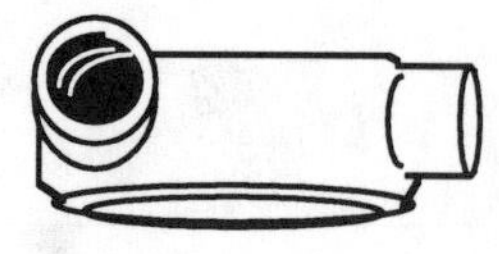

3. **I**

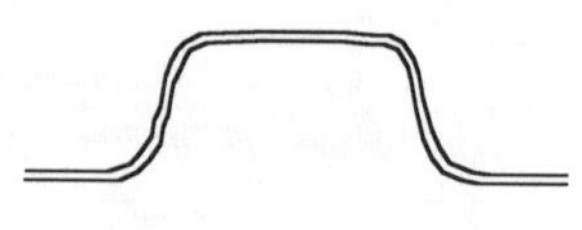

4. **O**

5. **D**

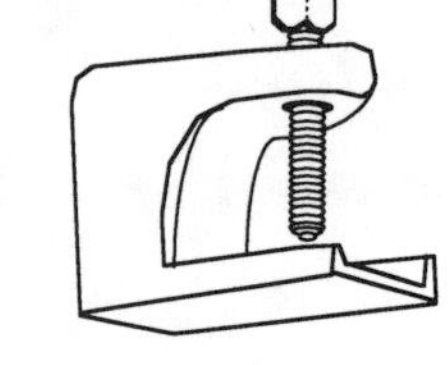

6. **C**

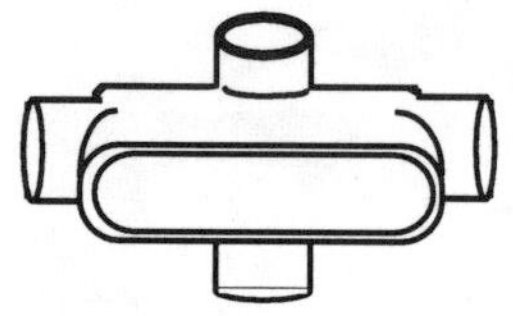

7. **L**

8. **B**

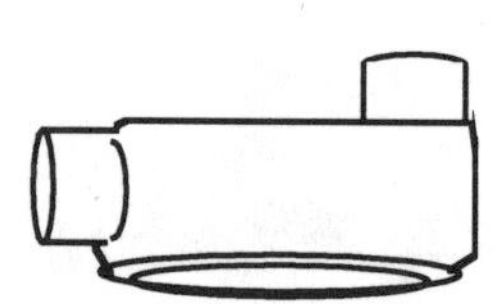

9. **F**

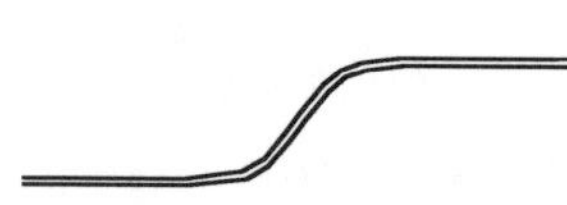

10. **J**

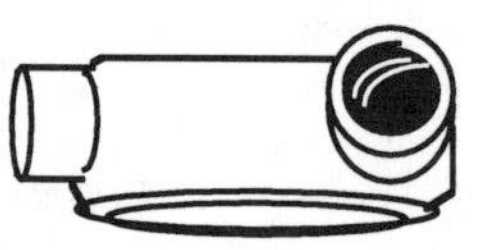

11. **P**

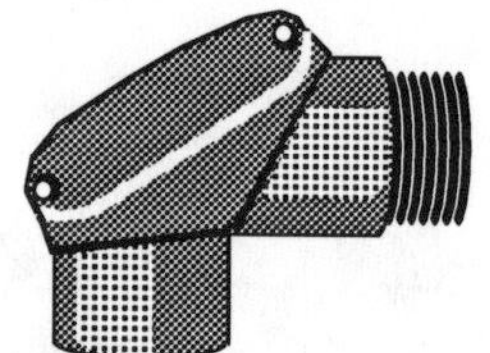

12. **G**

13. **N**

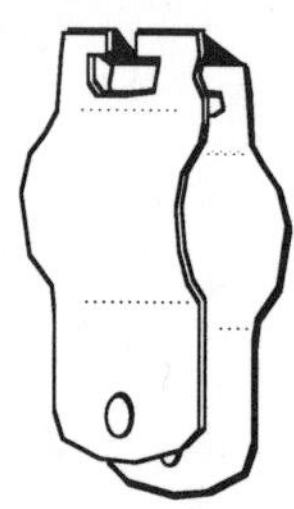

14. **E**

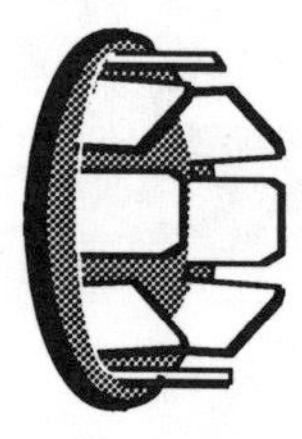

15. **H**

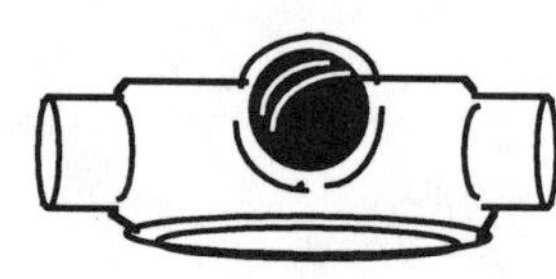

16. **A**

• *Choose a letter ( ) and fill in the blank above:*

(A) T conduit body
(B) plaster ring
(C) beam clamp
(D) back-to-back bend
(E) conduit hanger
(F) LB conduit body
(G) pulling 90° elbow
(H) knockout blank
(I) LL conduit body
(J) dog leg bend
(K) armored cable connector
(L) X conduit body
(M) cable sill plate
(N) reducing bushing
(O) saddle bend
(P) LR conduit body

JOURNEYMAN WIREMAN QUIZ #58
BLUEPRINT SYMBOLS

•Fill in the blank with the correct letter from choices below for the symbol

1. **J**

2. **M**

3. **E**

4. **Q**

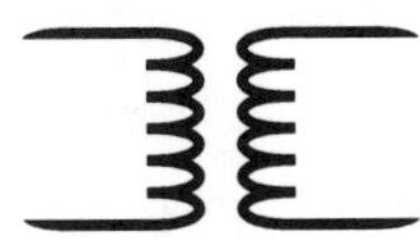

5. **N**

6. **Y**

S 4

7. **H**

8. **K**

TV

9. **F**

10. **S**

11. **A**

12. **C**

13. **B**

14. **P**

15. **U**

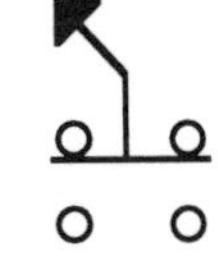

16. **X**

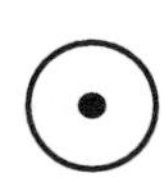

17. **Z**

18. **T**

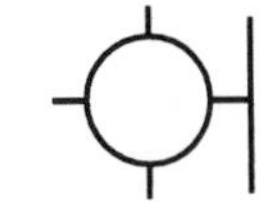

19. **D**

20. **R**

21. **L**

22. **G**

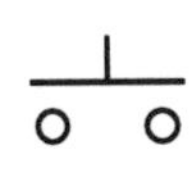

23. **W**

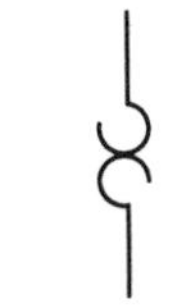

24. **I**

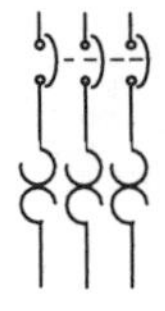

25. **O**

• *Choose a letter ( ) and fill in the blank above:*

(A) temperature actuated switch N.C.
(B) normally open foot switch
(C) phase
(D) wall bracket
(E) chime
(F) television outlet
(G) disconnect
(H) switch four-way
(I) thermal O.L.
(J) watthour meter
(K) liquid level switch N.C.
(L) SPDT double break
(M) mushroom head push button
(N) transformer
(O) CB with thermal O.L.
(P) delta
(Q) clock
(R) stop button N.C.
(S) N.O. contact
(T) timed contact N.O.T.C.
(U) limit switch N.O.
(W) start button N.O.
(X) selector switch two-position
(Y) N.C. contact
(Z) floor outlet

JOURNEYMAN QUIZ #59
TOOL IDENTIFICATION QUIZ

•Fill in the blank with the correct letter from choices below for the tool shown:

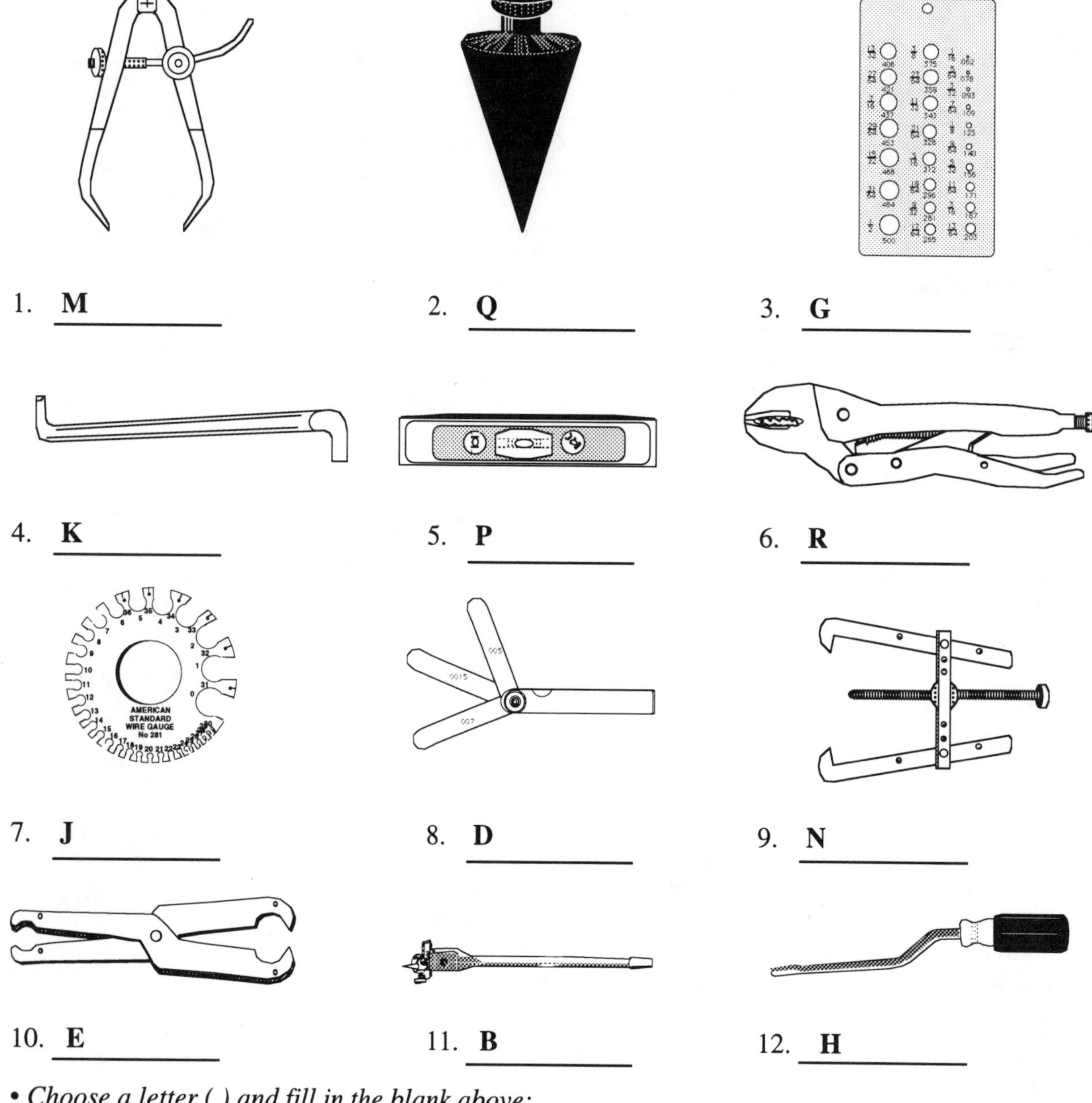

• Choose a letter ( ) and fill in the blank above:

(A) scratch awl
(B) expansion bit
(C) pump pliers
(D) feeler gauge
(E) fuse puller
(F) pipe holder
(G) drill bit gauge
(H) rotary screwdriver
(I) hole saw
(J) wire gauge
(K) offset screwdriver
(L) scribe
(M) caliper
(N) gear puller
(O) depth checker
(P) torpedo level
(Q) plumb bob
(R) locking pliers

JOURNEYMAN WIREMAN QUIZ #60
WIRING METHODS QUIZ

• *Fill in the blank with the correct letter for the wiring method or equipment:*

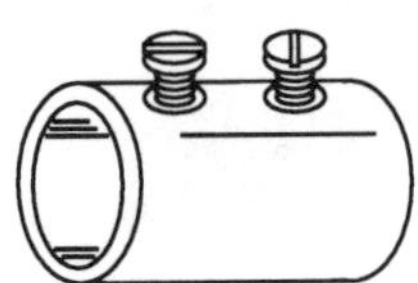
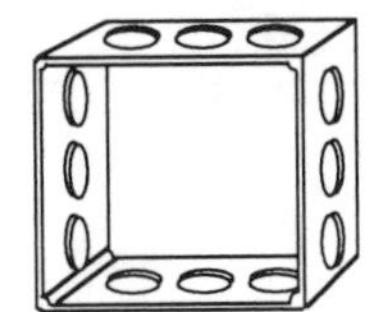
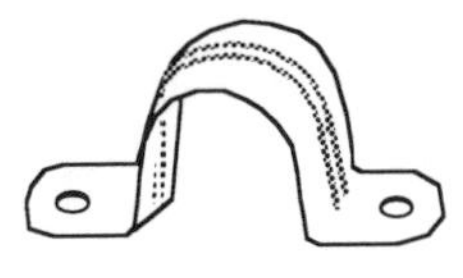

1. **N** 2. **M** 3. **P** 4. **I**

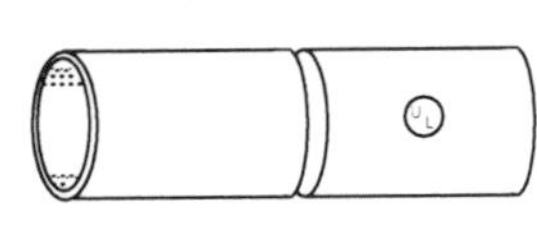

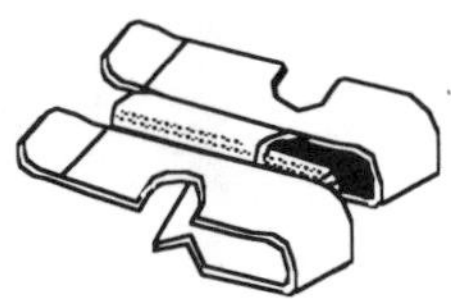

5. **J** 6. **L** 7. **G** 8. **F**

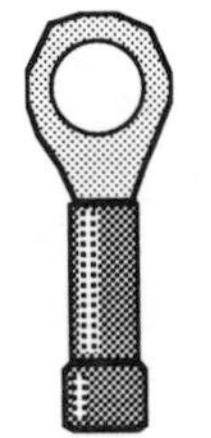
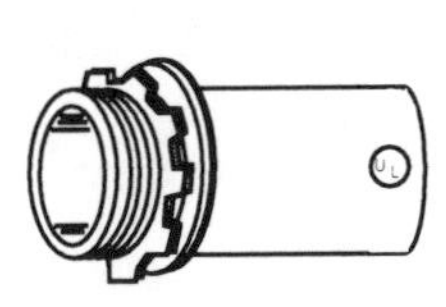
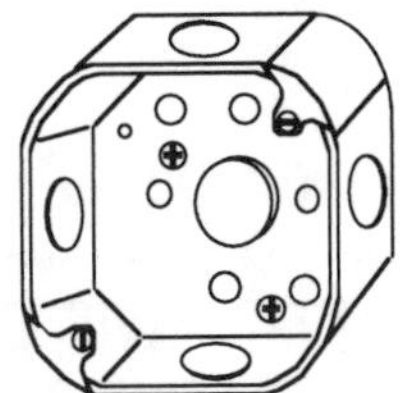
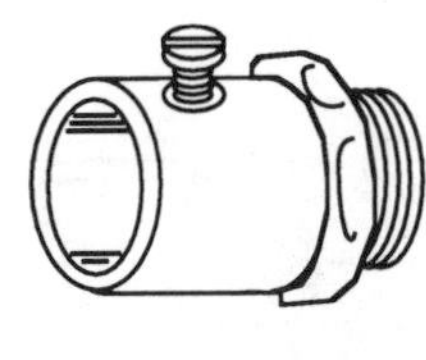

9. **A** 10. **E** 11. **B** 12. **C**

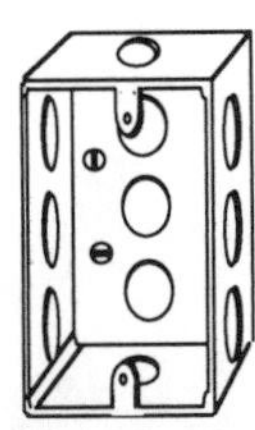

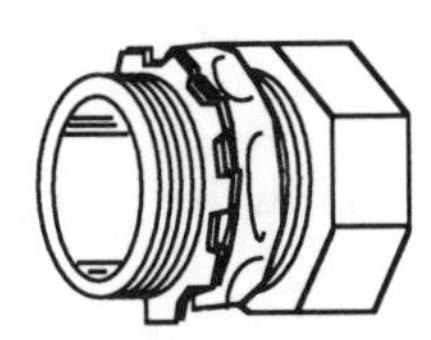
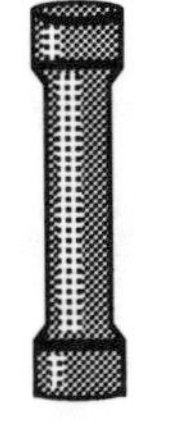

13. **O** 14. **K** 15. **H** 16. **D**

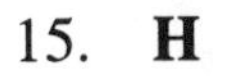

• *Choose a letter ( ) and fill in the blank above:*

(A) ring tongue terminal
(B) octagon box
(C) set screw type connector
(D) 50 amp receptacle
(E) indenter type connector
(F) grounding clip
(G) oval service cable strap
(H) butt connector
(I) sweeping elbow
(J) 30 amp receptacle
(K) compression type connector
(L) indenter type coupling
(M) square box
(N) set screw type coupling
(O) handy box
(P) pipe strap

JOURNEYMAN WIREMAN QUIZ #61
PRACTICAL QUESTIONS

•Circle the correct answer:

1. If the Line 1 fuse is blown and Line 2 fuse is okay, the test light that will be lit is # ____.

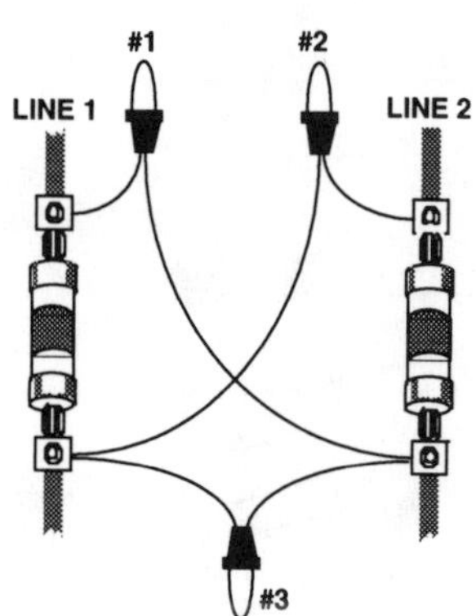

(a) #1 **(b) #2** **(c) #3** **(d) #1 and #2**

2. A recommended safe distance between the foot of an extension ladder and the wall that the ladder is placed against is _____.

(a) 1/2 the length of the extended ladder
(b) no more than 3 feet
(c) 1/4 the length of the extended ladder
(d) 5 feet for ladders less than 16 feet

3. Which wires are connected together so the switch will control both lights?

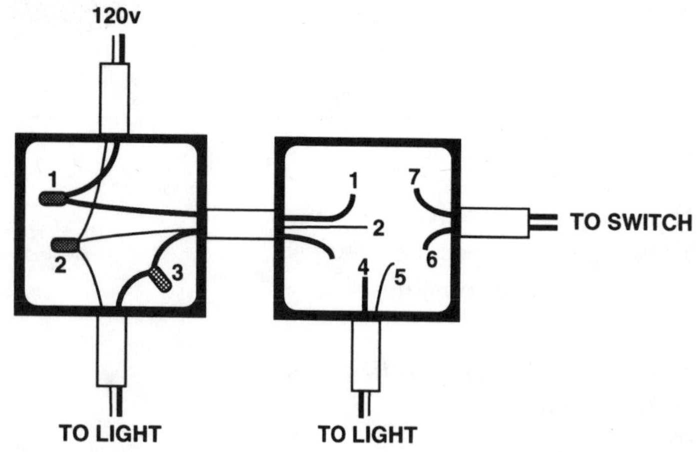

(a) 1 to 5, 2 to both 6 and 7, 3 to 4

(b) 1 to 7, 2 to both 5 and 6, 3 to 4

(c) 1 to 4, 2 to 6, 3 to both 5 and 7

(d) 1 to 6, 2 to 5, 3 to both 4 and 7

JOURNEYMAN WIREMAN QUIZ #62
PRACTICAL QUESTIONS

•Circle the correct answer:

1. A neon test light is connected across the terminals of a single-pole switch to a flourescent light on a 120 volt branch circuit. The test light will be lit when ____.

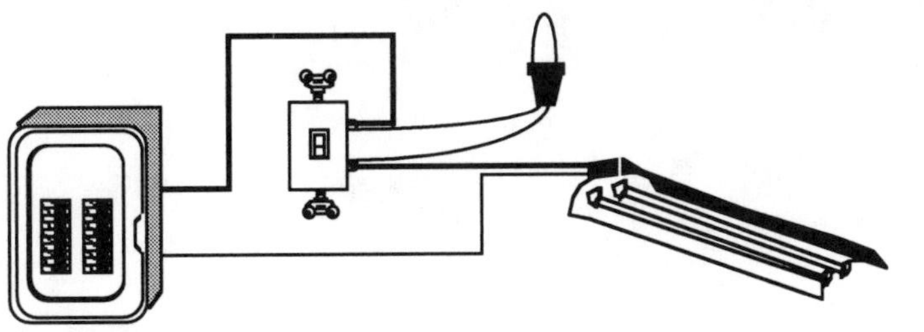

(a) the switch is turned on
(b) the switch is turned off
(c) both (a) and (b)
(d) neither (a) nor (b)

2.

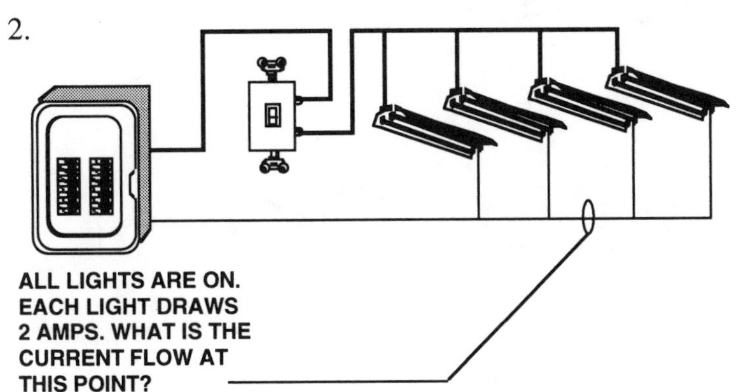

(a) 2 amps
(b) 4 amps
(c) 6 amps
(d) 8 amps

3. The meter will read ____ volts.

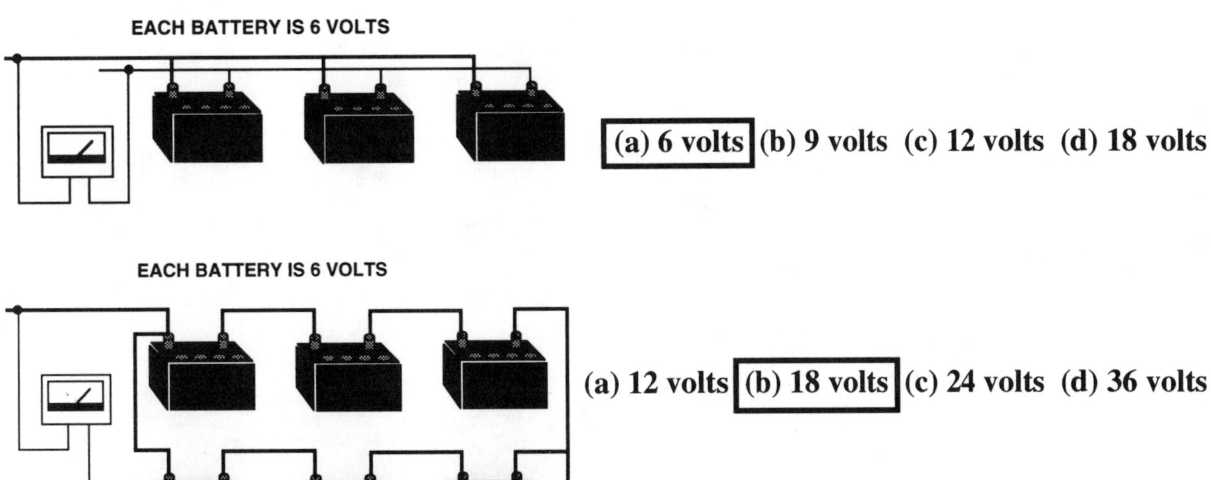

(a) 6 volts (b) 9 volts (c) 12 volts (d) 18 volts

(a) 12 volts (b) 18 volts (c) 24 volts (d) 36 volts

ANSWERS

EXAM #1

CLOSED BOOK

3 HOURS

PARTS BOARD EXAM #1

• *Fill in the blank with the correct letter shown below to identify the part.*

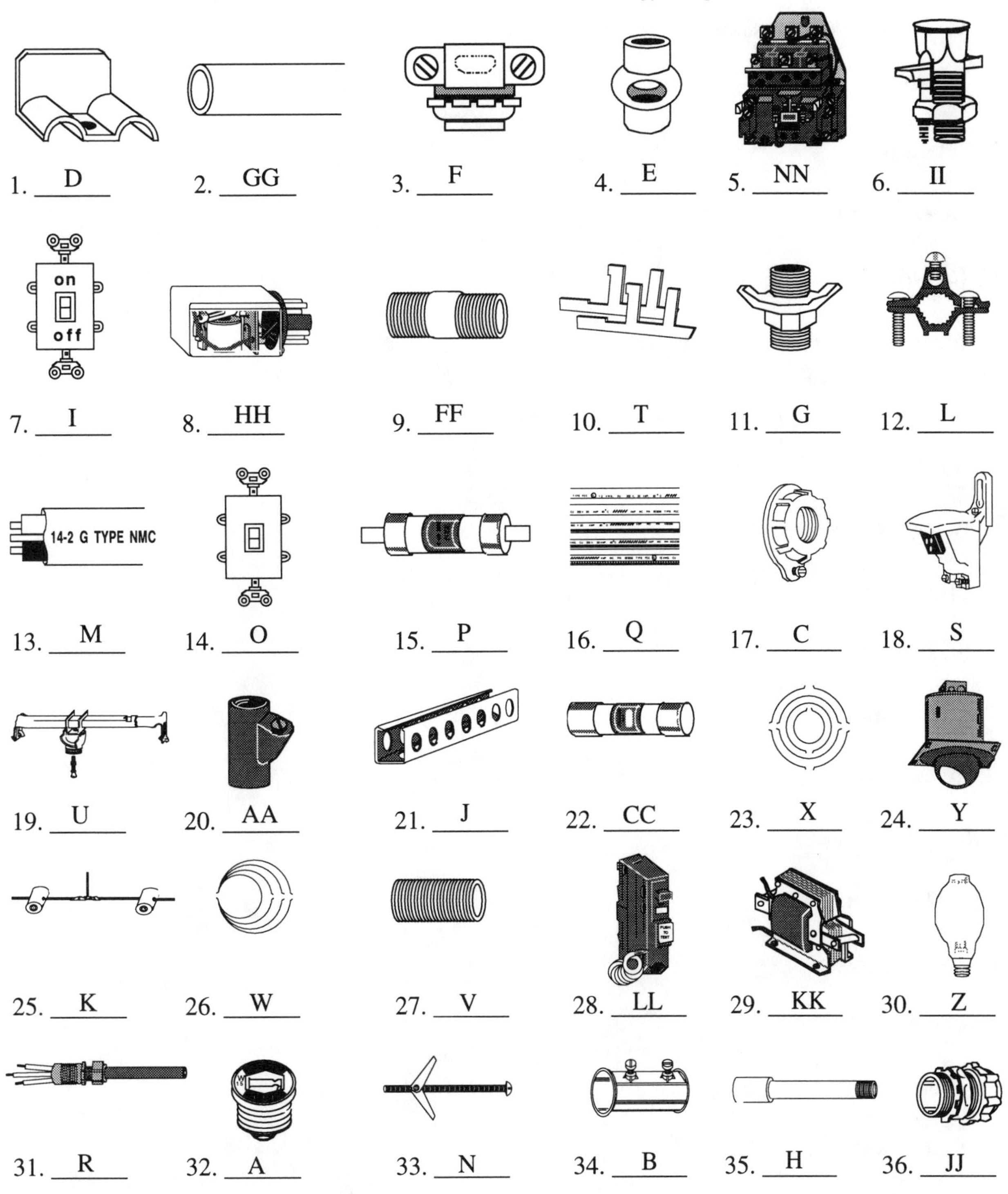

A. plug fuse B. EMT coupling C. grounding bushing D. cable clamp E. hickey F. two-screw connector G. fixture stud H. rigid conduit I. double-pole switch J. kindorf K. open wiring L. ground clamp M. Romex N. toggle bolt O. four-way switch P. knife fuse Q. FCC R. MI cable S. service head T. madison strap U. bar hanger V. close nipple W. eccentric KO X. concentric KO Y. bull's light Z. mercury lamp AA. seal off BB. GFCI receptacle CC. cartridge fuse DD. 3-way switch EE. BX cable FF. short nipple GG. EMT HH. relay II. split-bolt connector JJ. box connector KK. solenoid LL. GFCI breaker MM. pressure switch NN. motor starter

OHMS LAW - THEORY EXAM #1

1. **(c) increases the resistance**
2. **(d) 12Ω**
3. **(c) ohms**
4. **(c) current flows**
5. **(a) equal current flow**
6. **(b) reduced to the simplest form**
7. **(c) 1/4 as much**
8. **(c) effective difference**
9. **(c) ease of voltage variation**
10. **(a) sum**
11. **(c) shorter life**
12. **(a) 5 amps**
13. **(b) current**
14. **(a) real power**
15. **(c) 220 watt**
16. **(d) increases**
17. **(c) 6000 watt**
18. **(c) current develops heat**
19. **(b) 420 watt**
20. **(b) current**

CODE BOOK QUESTIONS EXAM #1

1. **(a) 80%   384-16c**
2. **(a) 6"   Table 300-5**
3. **(a) grounded   200-1**
4. **(a) 6 feet   210-52**
5. **(b) #18   402-6**
6. **(d) 8 feet   250-83c3**
7. **(a) 6"   300-14**
8. **(a) #10   110-14 ex.**
9. **(c) freedom from hazard   90-1b**
10. **(c) #1   310-4**
11. **(a) metal underground water pipe   250-81a**
12. **(a) 1/8 hp   422-21a**
13. **(b) 1/8"   370-21**
14. **(b) 10 feet   230-24b**
15. **(d) 20 feet   680-6a3**
16. **(c) #14 copper   Table 310-5**
17. **(b) four   348-10**
18. **(b) 2 cubic inch   Table 370-16b**
19. **(b) immediately   305-3d**
20. **(c) cover   370-25**

ANSWERS

EXAM #2

CLOSED BOOK

3 HOURS

PARTS BOARD EXAM #2

• *Fill in the blank with the correct letter shown below to identify the part.*

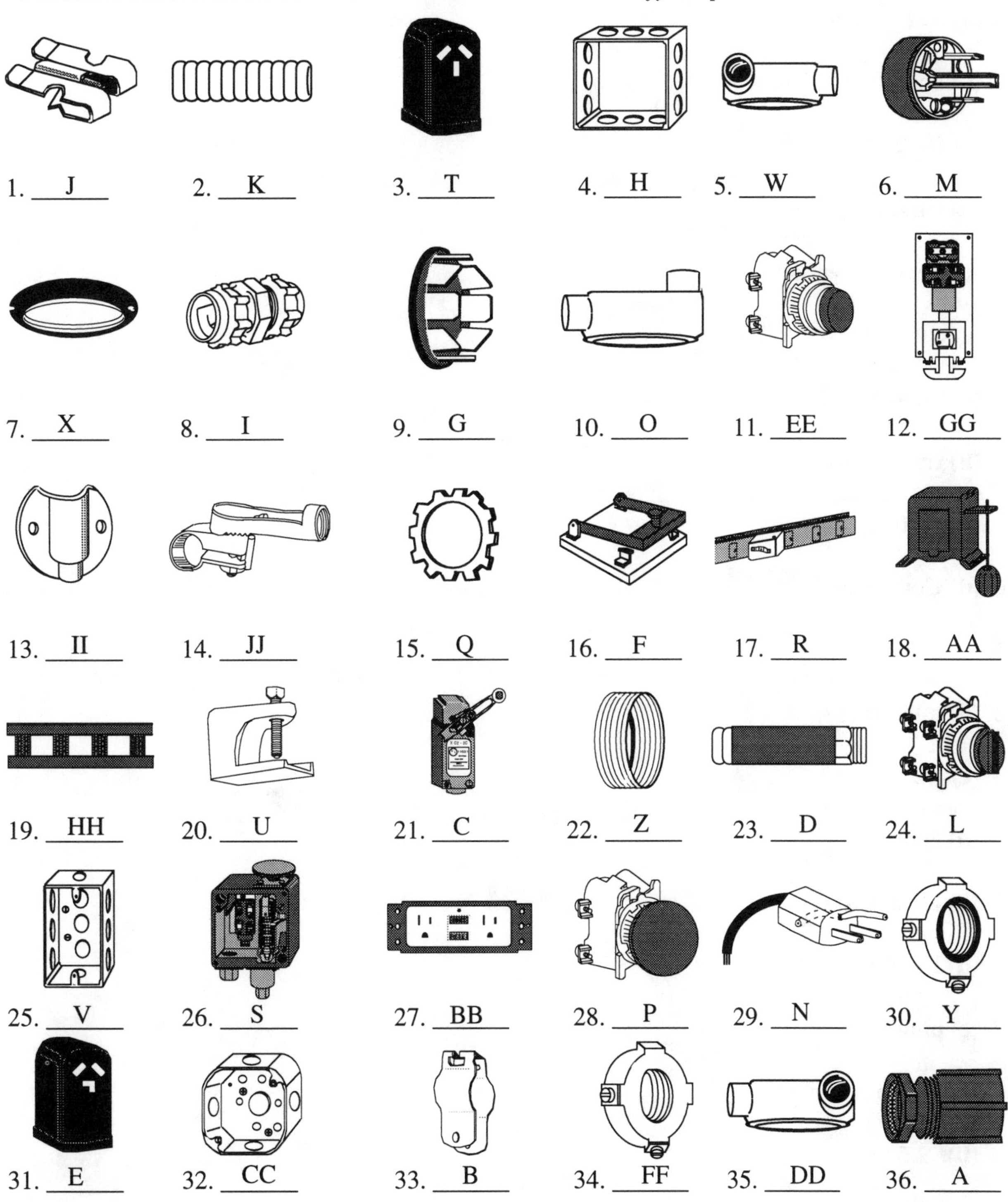

A. erickson B. conduit hanger C. limit switch D. liquid tight flex E. 30 amp receptacle F. knife switch G. knockout blank H. square box I. EMT coupling J. clip clamp K. BX L. selector switch M. attachment plug N. self-restoring plug O. LB P. stop button Q. locknut R. busway S. pressure switch T. 50 amp receptacle U. beam clamp V. handy box W. LL X. plaster ring Y. insulated throat bushing Z. reducing bushing AA. float switch BB. GFCI receptacle CC. octagon box DD. LR EE. start button FF. noninsulated throat bushing GG. timer HH. cable tray II. cable sill plate JJ. pipe ground connecter KK. solenoid LL. GFCI breaker MM. pressure switch

OHMS LAW - THEORY EXAM #2

1. **(a) 3 Ω**
2. **(b) good low resistance**
3. **(c) 10 Ω**
4. **(d) smallest particle**
5. **(c) electro chemistry**
6. **(c) current**
7. **(c) I and II only**
8. **(b) transformer**
9. **(d) rectifier**
10. **(d) Conductance**
11. **(c) 6 volts**
12. **(d) all of these**
13. **(c) 1/2**
14. **(d) current**
15. **(b) always less**
16. **(d) I and III**
17. **(c) phase**
18. **(d) DC to AC**
19. **(b) 2.5 Ω**
20. **(b) resistance**

CODE BOOK QUESTIONS EXAM #2

1. **(d) 10 feet 210-52h**
2. **(b) ungrounded 380-2a**
3. **(a) #4 200-6b 310-12a**
4. **(c) 86°F Table 310-16**
5. **(a) accidentally energized at a higher potential than ground 250-1 FPN**
6. **(c) grounded 110-16a**
7. **(c) adjacent to the basin 210-52d**
8. **(d) 50 pounds 410-16a**
9. **(a) yes, if insulated for the maximum voltage of any conductor 300-3c1**
10. **(a) 35 pounds 422-18**
11. **(c) 15 feet 230-24b**
12. **(a) accessible 250-112**
13. **(b) shall 90-5**
14. **(b) Festoon 225-6 definition**
15. **(b) 600 volts 110-30**
16. **(b) it is free of shorts and unintentional grounds 110-7**
17. **(d) I, II or III 110-13a**
18. **(d) I through V 90-1b**
19. **(c) 42 384-15**
20. **(d) I, II and III 250-51**